Math Made Visual

Creating Images for Understanding Mathematics

© 2006 by
The Mathematical Association of America (Incorporated)

Library of Congress Control Number 2005937269
ISBN 0-88385-746-4

Printed in the United States of America

Current Printing (last digit):
10 9 8 7 6 5 4 3 2 1

Math Made Visual

Creating Images for Understanding Mathematics

Claudi Alsina

Universitat Politècnica de Catalunya

and

Roger B. Nelsen

Lewis & Clark College

Published and Distributed by
The Mathematical Association of America

CLASSROOM RESOURCE MATERIALS

Classroom Resource Materials is intended to provide supplementary classroom material for students—laboratory exercises, projects, historical information, textbooks with unusual approaches for presenting mathematical ideas, career information, etc.

101 Careers in Mathematics, 2nd edition edited by Andrew Sterrett

Archimedes: What Did He Do Besides Cry Eureka?, Sherman Stein

Calculus Mysteries and Thrillers, R. Grant Woods

Combinatorics: A Problem Oriented Approach, Daniel A. Marcus

Conjecture and Proof, Miklós Laczkovich

A Course in Mathematical Modeling, Douglas Mooney and Randall Swift

Cryptological Mathematics, Robert Edward Lewand

Elementary Mathematical Models, Dan Kalman

Environmental Mathematics in the Classroom, edited by B. A. Fusaro and P. C. Kenschaft

Essentials of Mathematics, Margie Hale

Exploratory Examples for Real Analysis, Joanne E. Snow and Kirk E. Weller

Fourier Series, Rajendra Bhatia

Geometry From Africa: Mathematical and Educational Explorations, Paulus Gerdes

Historical Modules for the Teaching and Learning of Mathematics (CD), edited by Victor Katz and Karen Dee Michalowicz

Identification Numbers and Check Digit Schemes, Joseph Kirtland

Interdisciplinary Lively Application Projects, edited by Chris Arney

Inverse Problems: Activities for Undergraduates, Charles W. Groetsch

Laboratory Experiences in Group Theory, Ellen Maycock Parker

Learn from the Masters, Frank Swetz, John Fauvel, Otto Bekken, Bengt Johansson, and Victor Katz

Mathematical Connections: A Companion for Teachers and Others, Al Cuoco

Mathematical Evolutions, edited by Abe Shenitzer and John Stillwell

Mathematical Modeling in the Environment, Charles Hadlock

Mathematics for Business Decisions Part 1: Probability and Simulation (electronic textbook), Richard B. Thompson and Christopher G. Lamoureux

Mathematics for Business Decisions Part 2: Calculus and Optimization (electronic textbook), Richard B. Thompson and Christopher G. Lamoureux

Math Made Visual: Creating Images for Understanding Mathematics, Claudi Alsina and Roger B. Nelsen

Ordinary Differential Equations: A Brief Eclectic Tour, David A. Sánchez

Oval Track and Other Permutation Puzzles, John O. Kiltinen

A Primer of Abstract Mathematics, Robert B. Ash

Proofs Without Words, Roger B. Nelsen

Proofs Without Words II, Roger B. Nelsen

A Radical Approach to Real Analysis, David M. Bressoud

Real Infinite Series, Daniel D. Bonar and Michael Khoury, Jr.

She Does Math!, edited by Marla Parker

Solve This: Math Activities for Students and Clubs, James S. Tanton

Student Manual for Mathematics for Business Decisions Part 1: Probability and Simulation, David Williamson, Marilou Mendel, Julie Tarr, and Deborah Yoklic

Student Manual for Mathematics for Business Decisions Part 2: Calculus and Optimization, David Williamson, Marilou Mendel, Julie Tarr, and Deborah Yoklic

Teaching Statistics Using Baseball, Jim Albert

Topology Now!, Robert Messer and Philip Straffin

Understanding our Quantitative World, Janet Andersen and Todd Swanson

Writing Projects for Mathematics Courses: Crushed Clowns, Cars, and Coffee to Go, Annalisa Crannell, Gavin LaRose, Thomas Ratliff, Elyn Rykken

MAA Service Center
P.O. Box 91112
Washington, DC 20090-1112
1-800-331-1MAA   FAX: 1-301-206-9789

Dedicated to

Professor Berthold Schweizer

*for all the years of mathematical
collaboration and friendship*

Introduction

"a dull proof can be supplemented by a geometric analogue so simple and beautiful that the truth of a theorem is almost seen at a glance"
— Martin Gardner

"Behold!"
— Bhāskara

Is it possible to create mathematical drawings that help students understand mathematical ideas, proofs and arguments? We are convinced that the answer is yes and our objective in this book is to show how some visualization techniques may be employed to produce pictures that have both mathematical and pedagogical interest.

Mathematical drawings related to proofs have been produced since antiquity in China, Arabia, Greece and India but only in the last thirty years has there been a growing interest in so-called "proofs without words." Hundreds of these have been published in *Mathematics Magazine* and *The College Mathematics Journal*, as well as in other journals, books and on the World Wide Web. Popularizing this genre was the motivation for the second author of this book in publishing the collections [Nelsen, 1993 and 2000].

The first author became interested in creating proofs without words some years ago and more recently began a systematic study on how to teach others to design such pictures. This led him to organize and present many workshops on the topic devoted to secondary and university teachers. Consequently, we decided to join forces and prepare this book, extending a mathematical collaboration that goes back many years.

Often times, a person encountering a "proof without words" may have the feeling that the pictures involved are the result of a serendipitous discovery or the consequence of exceptional ingenuity on the part of the picture's creator. The next several chapters show that behind most of the pictures "proving" mathematical relations are some well-understood methods to follow. As will be seen, a given mathematical idea or relation may have many different images that justify it, so in the end, depending on the teaching level or the objectives for producing the pictures, one can choose the best alternative.

Since our main objective in this publication is to present a methodology for producing mathematical visualizations, we have divided the book into three parts:

Part I: Visualizing mathematics by creating pictures;

Part II: Visualization in the classroom;

Part III: Hints and solutions to the challenges.

Part I consists of twenty short chapters. Each one describes a method to visualize some mathematical idea (a proof, a concept, an operation, ...) and several applications to concrete cases, explained in detail. At the end of each chapter there is a collection of challenges so the reader may practice the abilities acquired during the reading of the preceding sections.

In Part II, after a short visit to the history of mathematical drawings, we present some general pedagogical considerations concerning the development of visual thinking, practical approaches for making visualizations in the classroom and, in particular, the role that hands-on materials may play in this process.

Finally, in Part III hints or solutions to all challenges of Part I are presented or discussed. We end the book with a list of references and a complete index of all topics considered. We hope the index will help teachers find particular concrete visualizations that they may be looking for, or even to organize a sequence of visual representations on a given topic (e.g., triangles, trigonometry, quadric surfaces, ...).

The reader will note that we have not included a chapter devoted to the use of technology. We feel that many, if not most, of the ideas presented here are independent of technology, and as befit the circumstances one might want to produce images in various ways—using chalk on the blackboard, a carefully hand-drawn transparency, or an image produced by commercial software. Our focus is on the creation of images rather than various methods of presenting images.

Making good pictures in the support of mathematics is always a challenging activity. We hope that by working through the following chapters the reader will encounter some new ideas, opening new windows to mathematical and pedagogical creativity. We have been fascinated by the processes of making mathematical drawings and we want to share with others a glimpse of this fascination.

Special thanks to Rosa Navarro for her superb work in the preparation of the final text of the manuscript, to Amadeu Monreal and Jerónimo Buxareu for their help in the preparation of some drawings. Thanks too to Zaven Karian and the members of the editorial board of Classroom Resource Materials for their careful reading of an earlier draft of the book and for their many helpful suggestions. We would also like to thank Elaine Pedreira, Beverly Ruedi, and Don Albers of the MAA's publication staff for their expertise in preparing this book for publication. Finally, special thanks to many groups of teachers in Argentina, Spain, and the United States for their willingness to explore these ideas and techniques with us, and for encouraging us to work on this publication.

Claudi Alsina
Universitat Politècnica
de Catalunya
Barcelona, Spain

Roger B. Nelsen
Lewis & Clark College
Portland, Oregon

Contents

Contents

Part I

Visualizing Mathematics by Creating Pictures

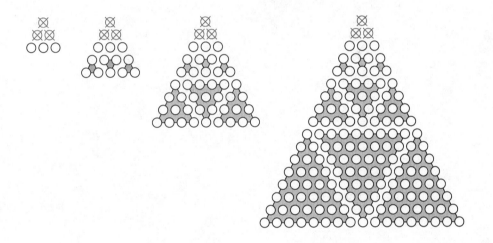

1

Representing Numbers
by Graphical Elements

In many problems concerning the natural numbers $(1, 2, \ldots)$, insight can be gained by representing the numbers by sets of objects. Since the particular choice of object is unimportant, we will usually use dots, squares, spheres, cubes, and other common easily drawn objects.

When one is faced with the task of verifying a statement concerning natural numbers (for example, showing that the sum of the first n odd numbers is n^2), a common approach is to use mathematical induction. However, such an analytical or algebraic approach rarely sheds light on *why* the statement is true. A geometric approach, wherein one can visualize the number relationship as a relationship between sets of objects, can often provide some understanding.

In this chapter we will illustrate two simple counting principles, both of which involve the representation of natural numbers by sets of objects. The principles are:

1. if you count the objects in a set two different ways, you will get the same result; and
2. if two sets are in one-to-one correspondence, then they have the same number of elements.

The first principle has been called the *Fubini principle* [Stein, 1979], after the theorem in multivariate calculus concerning exchanging the order of integration in iterated integrals. We call the second the *Cantor principle,* after Georg Cantor (1845–1918), who used it extensively in his investigations into the cardinality of infinite sets. We now illustrate the two principles. [Note: The two principles are actually equivalent.]

1.1 Sums of odd integers

Let's establish the statement about sums of odd numbers mentioned above, i.e., $1 + 3 + 5 + \cdots + (2n - 1) = n^2$. In Figure 1.1, we can count the dots in two ways, by multiplying

the number or rows by the number of columns ($n \times n$), or by the number of dots in each
L-shaped region ($1 + 3 + 5 + \cdots + (2n - 1)$). By the Fubini principle, the two counts
must be the same, which verifies the result.

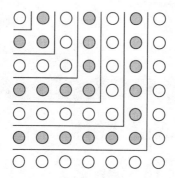

FIGURE **1.1**

Although we only illustrated the identity for the $n = 7$ case, the pattern clearly holds
for any natural number n.

In Figure 1.2, we see two sets of dots, the one on the right is simply a rearrangement
of the dots in the one on the left. It is easy to see a one to one correspondence between
the elements of the two sets (similarly colored dots correspond). Counting by rows in the
set on the left, we have $1 + 3 + 5 + \cdots + (2n - 1)$ dots, n^2 in the set on the right, and
the Cantor principle establishes the result.

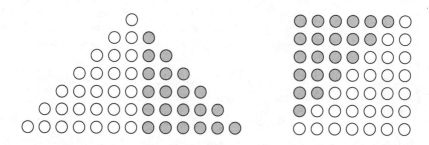

FIGURE **1.2**

1.2   Sums of integers

We can also use the two principles to establish the classical formula for the sum of the
first n natural numbers: $1 + 2 + \cdots + n = n(n + 1)/2$. If we adjoin a column of n dots to
the left side of the array in Figure 1.1, we obtain the array in Figure 1.3. Counting dots
by the L-shaped regions yields $2 + 4 + \cdots + 2n$, while multiplying the number of rows
by the number of columns yields $n(n + 1)$, hence the Fubini principle yields the desired
result (after division by 2).

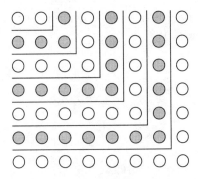

FIGURE 1.3

Alternatively, we can take two copies of $1 + 2 + \cdots + n$ and rearrange the dots, as shown in Figure 1.4. The set on the left has $2(1 + 2 + \cdots + n)$ dots, while that on the right has $n^2 + n$ dots. The Cantor principle (and division by 2 again) yields the desired result [Farlow, 1995].

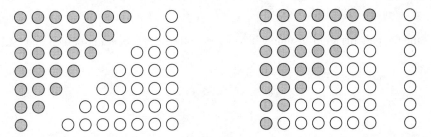

FIGURE 1.4

The arrangement of $1 + 2 + \cdots + n$ dots into the shape of a triangle in the left side of Figure 1.4 explains why the sum $1 + 2 + \cdots + n = n(n + 1)/2$ is often called the nth *triangular number,* which we denote T_n.

1.3 Alternating sums of squares

Squares and triangular numbers are both examples of what are called *figurate numbers,* since they can be represented by arrangements of objects into geometric figures (such as squares and triangles). There are many lovely relations among the figurate numbers, one of which is the following. Consider alternating sums of squares:

$$
\begin{aligned}
1^2 - 2^2 &= -3 = -(1 + 2); \\
1^2 - 2^2 + 3^2 &= +6 = +(1 + 2 + 3); \\
1^2 - 2^2 + 3^2 - 4^2 &= -10 = -(1 + 2 + 3 + 4); \quad \text{etc.}
\end{aligned}
$$

The resulting sums are triangular numbers, and it appears that the general pattern is

$$
1^2 - 2^2 + 3^2 - \cdots + (-1)^{n+1} n^2 = (-1)^{n+1} T_n.
$$

We can illustrate this pattern with dots, using shading to distinguish the dots that disappear in these operations [Logothetti, 1987]:

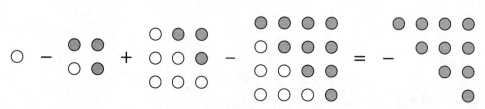

FIGURE **1.5**

1.4 Challenges

1.1 If $T_n = 1 + 2 + \cdots + n$ denotes the nth triangular number, show that

a. $T_{n-1} + T_n = n^2$,   b. $8T_n + 1 = (2n + 1)^2$,

c. $T_{2n} = 3T_n + T_{n-1}$,   d. $T_{2n+1} = 3T_n + T_{n+1}$,

e. $T_{3n+1} - T_n = (2n + 1)^2$,   f. $T_{n-1} + 6T_n + T_{n+1} = (2n + 1)^2$.

1.2 Find the patterns and illustrate with dots the following "ascending-descending" sums:

a.

$$1 + 2 + 1 = 2^2,$$
$$1 + 2 + 3 + 2 + 1 = 3^3,$$
$$1 + 2 + 3 + 4 + 3 + 2 + 1 = 4^2, \quad \text{etc.}$$

b.

$$1 + 3 + 1 = 1^2 + 2^2,$$
$$1 + 3 + 5 + 3 + 1 = 2^2 + 3^2,$$
$$1 + 3 + 5 + 7 + 5 + 3 + 1 = 3^2 + 4^2, \quad \text{etc.}$$

1.3 Show that the sum of consecutive powers of 9 is a triangular number:

$$1 + 9 = 10 = T_4,$$
$$1 + 9 + 81 = 91 = T_{13},$$
$$1 + 9 + 81 + 729 = 820 = T_{40}, \quad \text{etc.}$$

2

Representing Numbers by Lengths of Segments

A very natural way to represent a positive number a is to construct a line segment of length a. In this way many relationships between positive numbers may be illustrated with figures, and relationships among lengths of line segments in those figures.

Given two segments of lengths $a, b > 0$ and a segment of unit length, we describe representations of some basic quantities associated with a and b in Figure 2.1.

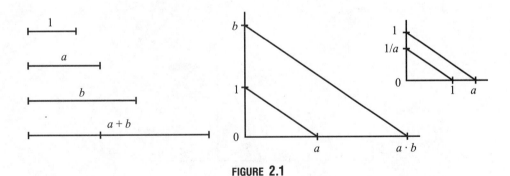

FIGURE 2.1

2.1 Inequalities among means

The best known and most common way to "average" two numbers a and b is the *arithmetic mean* $(a + b)/2$, which always lies between a and b. But there are other means. The geometric mean of two positive numbers a and b is $\sqrt{ab}$, which again lies between a and b. For example, the Weber-Fechner law in psychology states that perception varies as the logarithm of the stimulus. Consequently, it is the geometric mean of two stimuli that is perceived as the arithmetic mean of their respective perceptions.

How do the arithmetic mean and geometric mean compare? In Figure 2.2 we show that for $0 < a < b$, we have $a < \sqrt{ab} < (a + b)/2 < b$. Note that (i) a triangle inscribed in a semicircle is a right triangle, (ii) the altitude to the hypotenuse divides a right triangle into two smaller right triangles similar to the original, and (iii) ratios of corresponding sides of similar triangles are equal; hence $a/h = h/b$, or $h = \sqrt{ab}$. Noting that the longest perpendicular from a semicircle to its diameter is the radius (see Figure 2.2(b)) establishes the inequality [Gallant, 1977].

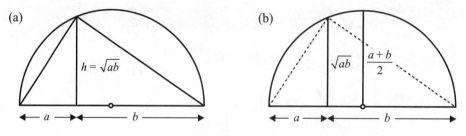

FIGURE 2.2

Another mean of interest is the *harmonic mean*: for positive numbers a and b it is given by $2ab/(a + b)$ and again lies between a and b. For example, if one drives D km at a speed a km/h, and returns D km at a speed b km/h, the average speed for the round trip is $2ab/(a+b)$ km/h. The harmonic mean is smaller than the geometric and arithmetic means for $0 < a < b$, as shown in Figure 2.3, a demonstration due to Pappus of Alexandria (circa A.D. 320) [Cusmariu, 1981]. Again, the inequalities result from comparisons of the lengths of sides in similar triangles.

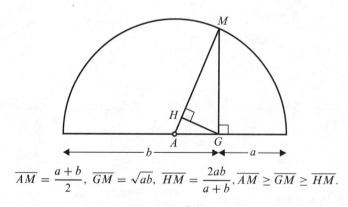

$$\overline{AM} = \frac{a+b}{2}, \quad \overline{GM} = \sqrt{ab}, \quad \overline{HM} = \frac{2ab}{a+b}, \overline{AM} \geq \overline{GM} \geq \overline{HM}.$$

FIGURE 2.3

The final mean we consider is the *root-mean square,* which for positive numbers a and b is given by $\sqrt{(a^2 + b^2)/2}$. For example, given two squares with side lengths a and b, the side of a square whose area is the arithmetic mean of a^2 and b^2 is the root-mean-square of a and b. The root-mean-square is larger than the three means previously considered,

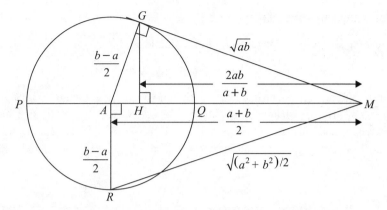

$$0 < a < b, \overline{QM} = a, \overline{PM} = b, \overline{HM} < \overline{GM} < \overline{AM} < \overline{RM}.$$

FIGURE 2.4

so that we have

$$0 < a < b \quad \text{implies} \quad a < \frac{2ab}{a+b} < \sqrt{ab} < \frac{a+b}{2} < \sqrt{\frac{a^2+b^2}{2}} < b,$$

which is shown in Figure 2.4 and follows from comparisons of lengths of sides in similar triangles and the Pythagorean theorem.

2.2 The mediant property

If a, b, c, d are positive numbers such that $a/b < c/d$, then the fraction $(a+b)/(b+d)$ formed by adding numerators and adding denominators is called the *mediant* of a/b and c/d, and it always lies between a/b and c/d, that is,

$$\frac{a}{b} < \frac{c}{d} \quad \text{implies} \quad \frac{a}{b} < \frac{a+c}{b+d} < \frac{c}{d}.$$

Representing a, b, c, d by line segments and interpreting the fractions as slopes of line segments leads to the following demonstration [Gibbs, 1990]:

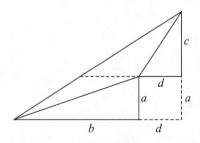

FIGURE 2.5

2.3   A Pythagorean inequality

In this section we present a picture-proof of a simple double inequality: For all $a, b > 0$,

$$\sqrt{a^2 + b^2} < a + b \leq \sqrt{2} \cdot \sqrt{a^2 + b^2}.$$

In Figure 2.6(a) we represent a and b by the legs of a right triangle with hypotenuse $c = \sqrt{a^2 + b^2}$. By the triangle inequality, $c < a + b$, which proves the first inequality. For the second, note that the side length $a + b$ of the square is less than or equal to the length $c\sqrt{2}$ of the diagonal of the square with side c. Figure 2.6(b) shows that when $a = b$, then $a + b = c\sqrt{2}$.

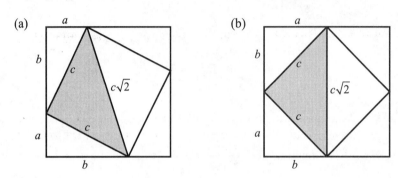

FIGURE 2.6

2.4   Trigonometric functions

The values of the six trigonometric functions of an acute angle θ can be represented by the lengths of line segments in three triangles, as seen in Figure 2.7 [Romaine, 1988]:

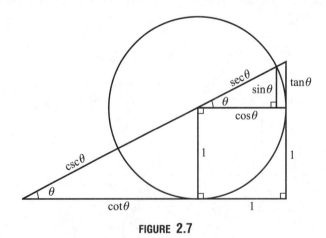

FIGURE 2.7

Note that Figure 2.7 also establishes the identity $(\tan \theta + 1)^2 + (\cot \theta + 1)^2 = (\sec \theta + \csc \theta)^2$.

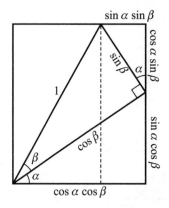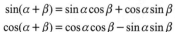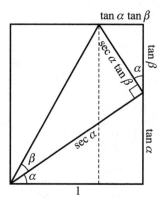

$$\sin(\alpha+\beta)=\sin\alpha\cos\beta+\cos\alpha\sin\beta$$
$$\cos(\alpha+\beta)=\cos\alpha\cos\beta-\sin\alpha\sin\beta$$

$$\tan(\alpha+\beta)=\frac{\tan\alpha+\tan\beta}{1-\tan\alpha\tan\beta}$$

FIGURE 2.8

Various basic trigonometric identities can be represented similarly by identifying line segments in certain triangles with trigonometric functions of appropriate angles. Figure 2.8 illustrates the addition formulas for the sine, cosine, and tangent functions of acute angles:

2.5 Numbers as function values

Given the graph $y = f(x)$ of a nonnegative function f, the vertical line segment connecting the points $(a, 0)$ and $(a, f(a))$ represents the number $f(a)$. As an illustration, we show that for $e \le a < b$, $a^b > b^a$. In Figure 2.9 [Gallant, 1991], we see that the slope of the line L_a (i.e., $\ln a/a$) is greater than the slope of the line L_b (i.e., $\ln b/b$), which then leads to the desired result.

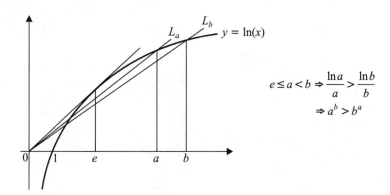

$$e \le a < b \Rightarrow \frac{\ln a}{a} > \frac{\ln b}{b}$$
$$\Rightarrow a^b > b^a$$

FIGURE 2.9

In particular, when $a = e$ and $b = \pi$ we have $e^\pi > \pi^e$.

2.6 Challenges

2.1 Use Figure 2.10 to illustrate the half-angle tangent formulas:

$$\tan\frac{\theta}{2} = \frac{\sin\theta}{1+\cos\theta} = \frac{1-\cos\theta}{\sin\theta}.$$

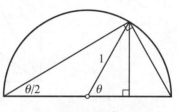

FIGURE **2.10**

2.2 Find illustrations similar to those in Figure 2.7 for the subtraction formulas for the sine, cosine, and tangent.

2.3 Illustrate the fact that if α, β, γ are positive angles such that $\alpha + \beta + \gamma = \pi/2$, then $\tan\alpha\,\tan\beta + \tan\beta\,\tan\gamma + \tan\gamma\,\tan\alpha = 1$. [Hint: The diagram in Figure 2.7 can be used.]

2.4 Show that $1 - \theta^2/2 \le \cos\theta$ for any θ in $[0, \pi/2]$.

2.5 Prove *Bernoulli's inequality*: for any $x \ge -1$ and any $a > 1$, $(1 + x)^a \ge 1 + ax$.

3

Representing Numbers by Areas of Plane Figures

Another very natural way to represent a (positive) number is by the area of a region in the plane. The simplest such regions are squares and rectangles, and calculus can be used to represent a number as the area under the graph of a function. Counting problems now become area computations, and inequalities between numbers can be established by showing that one region has a larger or smaller area than another.

3.1   Sums of integers revisited

In Chapter 1 we encountered several visual representations for the nth triangular number $T_n = 1 + 2 + \cdots + n$. If we use a square of unit area to represent the number 1, two such squares to represent 2, and so on, then the area of Figure 3.1(a) represents T_n. To compute the area, we bisect diagonally the right-most squares in each row as shown in Figure 3.1(b), and compute the areas of the resulting large unshaded triangle and the n smaller shaded triangles [Richards, 1984]:

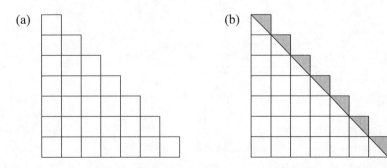

FIGURE **3.1**

Thus

$$T_n = 1 + 2 + \cdots + n = \frac{1}{2}n^2 + n \cdot \frac{1}{2} = \frac{n(n+1)}{2}.$$

Another way to evaluate T_n is to take two copies of the region in Figure 3.1(a) and then compute the area. Here $2T_n = n(n+1)$, and hence (once again) $T_n = \frac{1}{2}n(n+1)$ (see Figure 3.2).

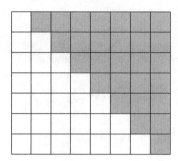

FIGURE **3.2**

3.2   The sum of terms in arithmetic progression

Since $1 + 2 + \cdots + n$ is the sum of n terms in arithmetic progression, perhaps the same ideas in the previous section will allow us to illustrate and evaluate the sum S of n numbers in a general arithmetic progression with first term a and common difference d:

$$S = a + (a + d) + (a + 2d) + \cdots + [a + (n-1)d].$$

Generalizing Figure 3.2 we have the following, which has been called the "organ-pipe" method for summing numbers in an arithmetic progression [Conway and Guy, 1996]:

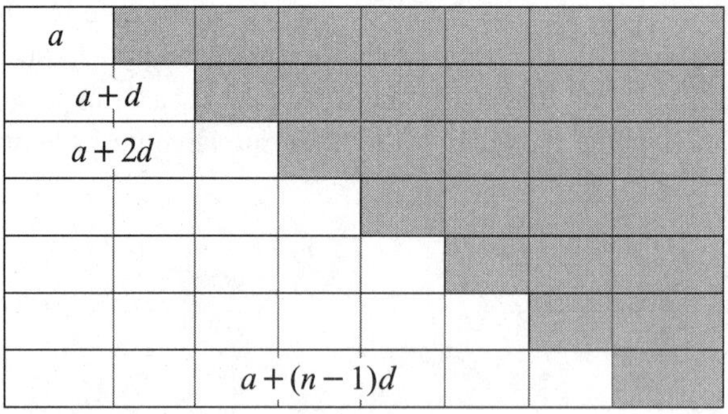

a						
$a+d$						
$a+2d$						
		$a+(n-1)d$				

FIGURE **3.3**

Hence $2S = n[a + a + (n-1)d]$, so that $S = (n/2)[2a + (n-1)d]$. The figure is certainly familiar to anyone who has worked with Cuisenaire rods. We will generalize in another direction by considering the sum of numbers in two-dimensional arithmetic progressions in the next chapter.

3.3 Fibonacci numbers

Leonardo Fibonacci (1175–1250) is probably best remembered for the sequence that bears his name. It arises as a solution to the following problem from his 1202 book *Liber abaci*: "How many pairs of rabbits can be produced from a single pair in a year if every month each pair begets a new pair which from the second month on becomes productive?" The monthly totals of rabbit pairs are 1, 1, 2, 3, 5, 8, 13 and so on, a sequence in which, after the first two terms, each term is the sum of the preceding two numbers. If F_n represents the nth Fibonacci number, then $F_1 = F_2 = 1$ and $F_n = F_{n-1} + F_{n-2}$ for $n \geq 3$.

Some of the prettiest identities for Fibonacci numbers concern sums of squares or sums of products of Fibonacci numbers. For example, $F_1^2 + F_2^2 + \cdots + F_n^2 = F_n F_{n+1}$, which can be illustrated as shown below (for $n = 6$) in Figure 3.4 [Brousseau, 1972]:

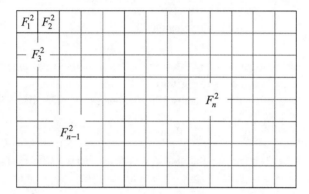

FIGURE **3.4**

Other identities can be illustrated similarly [Bicknell and Hoggatt, 1972]:

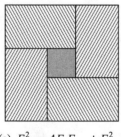 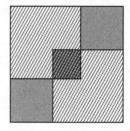

(a) $F_{n+1}^2 = 4F_n F_{n-1} + F_{n-2}^2$ (b) $F_{n+1}^2 = 2F_n^2 + 2F_{n-1}^2 - F_{n-2}^2$

FIGURE **3.5**

3.4 Some inequalities

The same picture used in Figure 3.5(a) can be employed to show that the sum of a positive number and its reciprocal is always at least 2, i.e., for any positive x, $x + (1/x) \geq 2$. Since the product of x and $1/x$ is 1, arrange four rectangles of these dimensions into a square, as shown in Figure 3.5(a). Since the area of each rectangle is 1, the area of the resulting square is at least 4, and thus the length $x + (1/x)$ of a side of the square must be at least 2.

This idea can be generalized to give another proof of the arithmetic mean-geometric mean inequality (which we encountered in the previous chapter) by letting a and b denote the dimensions of each of the four rectangles [Schattschneider, 1986]:

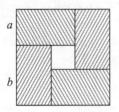

$$(a + b)^2 \geq 4ab$$

$$\therefore \frac{a + b}{2} \geq \sqrt{ab}$$

FIGURE **3.6**

In calculus, one application of the definite integral is the area under the graph of a function. Interpreting the natural logarithm of a positive number as the area under the graph of $y = 1/x$ over an interval leads to a demonstration of *Napier's inequality*,

$$\text{if } b > a > 0, \quad \text{then } \frac{1}{b} < \frac{\ln b - \ln a}{b - a} < \frac{1}{a},$$

as seen in Figure 3.7. Napier's inequality can be used to derive a familiar limit expression for the number e. Let $a = 1$ and $b = 1 + 1/n$, and take the limit as n approaches infinity:

$$\frac{n}{n + 1} \cdot \frac{1}{n} < \ln\left(1 + \frac{1}{n}\right) < 1 \cdot \frac{1}{n};$$

$$\frac{n}{n + 1} < n \ln\left(1 + \frac{1}{n}\right) < 1;$$

$$\lim_{n \to \infty} \ln\left(1 + \frac{1}{n}\right)^n = 1;$$

$$\therefore \lim_{n \to \infty} \left(1 + \frac{1}{n}\right)^n = e.$$

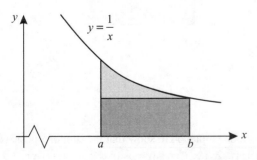

$$\frac{1}{b}(b - a) < \int_a^b \frac{1}{x} dx < \frac{1}{a}(b - a)$$

$$\therefore \frac{1}{b} < \frac{\ln b - \ln a}{b - a} < \frac{1}{a}$$

FIGURE **3.7**

3.5   Sums of squares

Here's an area illustration of the identity $1^2 + 2^2 + 3^2 + \cdots n^2 = n(n + 1)(2n + 1)/6$, established by showing that $3(1^2 + 2^2 + 3^2 + \cdots n^2)$ is equal in area to a rectangle whose dimensions are $2n + 1$ by $1 + 2 + \cdots + n$ [Gardner, 1973]:

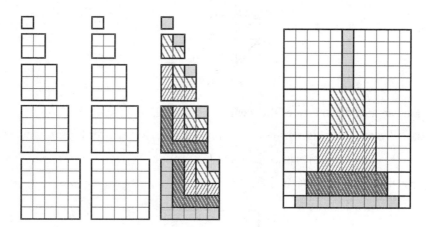

FIGURE **3.8**

3.6   Sums of cubes

We can use plane areas to illustrate an identity for cubes of integers by representing a cube such as n^3 as n copies of n^2 for an integer n. For example, Figure 3.9 illustrates the identity

$$1^3 + 2^3 + 3^3 + \cdots + n^3 = (1 + 2 + 3 + \cdots + n)^2.$$

In the figure, where two squares overlap in a smaller square, there is always an adjacent empty smaller square of the same area [Golomb, 1965].

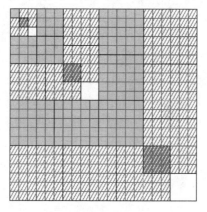

FIGURE **3.9**

3.7   Challenges

3.1 What identity for integers is represented by Figure 3.10?

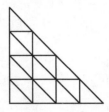

FIGURE **3.10**

3.2 Find an area illustration of the identity $F_{n+1}^2 = F_n^2 + F_{n-1}^2 + 2F_{n-1}F_n$.

3.3 Use Figure 3.11 to demonstrate the following two identities for Fibonacci numbers:

   a. $F_{n+1}^2 = 4F_{n-1}^2 + 4F_{n-1}F_{n-2} + F_{n-2}^2$,
   b. $F_{n+1}^2 = 4F_n^2 - 4F_{n-1}F_{n-2} - 3F_{n-2}^2$.

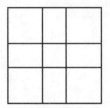

FIGURE **3.11**

3.4 Change the dimensions of the rectangles in Figure 3.6 to $a/(a+b)$ by $b/(a+b)$ to obtain a proof of the harmonic mean-geometric mean inequality.

3.5 Use areas to illustrate the "completing the square" identity:

$$x^2 + ax = (x + a/2)^2 - (a/2)^2.$$

3.6 Use areas to illustrate the identity

$$ax - by = \frac{1}{2}(a + b)(x - y) + \frac{1}{2}(a - b)(x + y)$$

   for $a > b > 0$ and $x > y > 0$.

3.7 Show that the product of four consecutive positive integers is always one less than a perfect square (e.g., $3 \cdot 4 \cdot 5 \cdot 6 = 360 = 19^2 - 1$).

3.8 Find an area illustration for the following: if p and q are positive, then

$$\int_0^1 \left(t^{p/q} + t^{q/p} \right) dt = 1.$$

   [Hint: if $y = x^{p/q}$, then $x = y^{q/p}$.]

4

Representing Numbers
by Volumes of Objects

In this chapter we represent a positive number by the volume of an object. In the simplest cases we can represent a product of three positive integers by the volume of a rectangular solid. We also can represent an integer by a collection of unit cubes, and establish identities by computing the volume. In many instances we may need to alter or rearrange the parts of an object before computing the volume.

4.1   From two dimensions to three

The reader is probably well acquainted with the following area representation of the familiar formula for factoring the difference of two squares:

$$a^2 - b^2 = (a - b)(a + b).$$

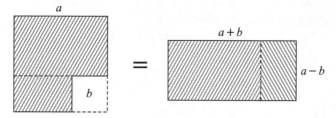

FIGURE 4.1

Using volumes, we have an analogous representation for factoring the difference of two cubes:

$$a^3 - b^3 = (a - b)(a^2 + ab + b^2).$$

19

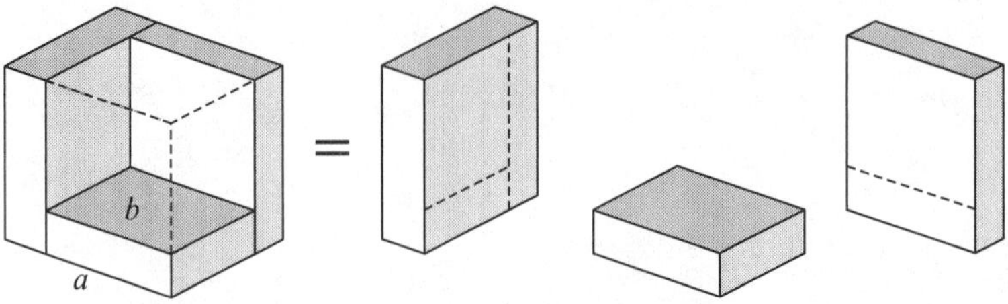

$$a^3 - b^3 \;\;=\;\; a^2(a-b) \;\;+\;\; b^2(a-b) \;\;+\;\; ab(a-b)$$

FIGURE **4.2**

In Section 3.3 we saw how to illustrate identities for the Fibonacci numbers using area (also see Challenge 3.2). In Figure 4.3 we have a volume representation of the identity

$$F_{n+1}^3 = F_n^3 + F_{n-1}^3 + 3F_{n-1}F_nF_{n+1}.$$

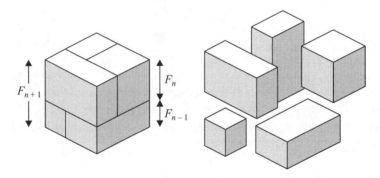

FIGURE **4.3**

4.2 Sums of squares of integers revisited

In Section 3.1 we saw a visual representation for the nth triangular number $T_n = 1 + 2 + \cdots + n$ as the area of a region composed of a collection of unit squares, and used areas of triangles to establish the formula for T_n (Figure 3.1). Analogously we can represent an integral square k^2 as a collection of k^2 unit cubes, and compute the sum $1^2 + 2^2 + \cdots n^2$ via volumes as illustrated in Figure 4.4:

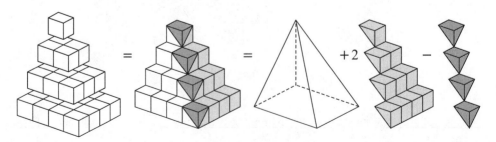

FIGURE **4.4**

Thus $1^2 + 2^2 + \cdots + n^2 = \frac{1}{3}n^2 \cdot n + 2 \cdot \frac{n(n+1)}{2} \cdot \frac{1}{2} - n \cdot \frac{1}{3} = \frac{n(n+1)(2n+1)}{6}$.

In the calculation we used the formula for the volume of a pyramid (1/3 the area of the base times the height) and $n(n+1)/2$ for the sum $1 + 2 + \cdots + n$.

4.3 Sums of triangular numbers

In a similar fashion we can find a formula for the sum of the first n triangular numbers. After stacking sets of unit cubes representing the triangular numbers, we "slice" off small pyramids (shaded gray in Figure 4.5) and place each small pyramid on the top of the cube from which it came. The result is a larger pyramid minus some smaller pyramids along one edge of the base:

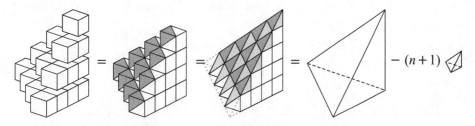

FIGURE 4.5

Thus $T_1 + T_2 + \cdots + T_n = \frac{1}{6}(n+1)^3 - (n+1) \cdot \frac{1}{6} = \frac{n(n+1)(n+2)}{6}$.

4.4 A double sum

In Section 3.1 we saw that "slicing" the region could be avoided by "duplicating" the region before the area computation (see Figures 3.1 and 3.2). We can use a similar technique in three dimensions to show that

$$\sum_{i=1}^{n} \sum_{j=1}^{n} (i + j - 1) = n^3 :$$

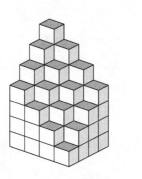 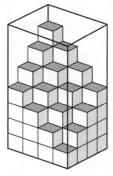

FIGURE 4.6

If S denotes the double sum in question, then $2S = 2n^3$, from which the identity follows. This technique of making multiple copies of the object will be explored further in Chapter 11.

4.5 Challenges

4.1 Find a volume representation for $a^3 + b^3 = (a + b)(a^2 - ab + b^2)$.

4.2 Show that the result in Section 4.4 generalizes:

$$\sum_{i=1}^{m}\sum_{j=1}^{n}\left[a + (i - 1)b + (j - 1)c\right] = \frac{mn}{2}\left[2a + (m - 1)b + (n - 1)c\right].$$

In other words, the sum of the terms in a two-dimensional arithmetic progression is one-half the number of terms times the sum of the first $[(i, j) = (1, 1)]$ and last $[(i, j) = (m, n)]$ terms.

4.3 Use Figure 4.7 to establish the following identities:

$$1 + 2 = 3,$$
$$4 + 5 + 6 = 7 + 8,$$
$$9 + 10 + 11 + 12 = 13 + 14 + 15, \quad \text{etc.}$$

FIGURE **4.7**

5

Identifying Key Elements

Mathematical pictures go beyond usual artistic representations because they often contain a great deal of information. They are sophisticated creatures that contain symbols as well as lines, angles, projections, measures, etc. In this chapter we illustrate the technique of introducing special "marks" in the pictures to identify relevant parts: equality of segments, equality of angles, repetitions, similar or congruent subsets, etc. In many cases appropriate identification of key elements readily yields a proof of the desired result. This is also the case in Euclidean geometry where, using straightedge and compass, one must construct figures using a collection of related elements (sides, angles, bisectors,...). The procedure becomes a process of identifying how the data determine the unknown parts. In making mathematical drawings ... details matter!

5.1 On the angle bisectors of a convex quadrilateral

In a triangle the three angle bisectors meet at the incenter. What happens in a convex quadrilateral? The following result gives the complete answer, and the proof is based on a simple picture in which all the relevant angles are identified.

> Given any convex quadrilateral, if its four angle bisectors determine a new quadrilateral, then the new quadrilateral is cyclic (i.e. it can be inscribed in a circle).

We can make a simple picture including the basic elements described in the above statement, and we mark on it the key angles (see Figure 5.1). But the angles a, b, x, y, z, t satisfy

$$a + x + y = \pi, \quad b + z + t = \pi, \quad 2x + 2y + 2z + 2t = 2\pi.$$

Consequently

$$a + b = (\pi - x - y) + (\pi - z - t) = 2\pi - (x + y + z + t) = \pi,$$

i.e., the new quadrilateral must be inscribable in a circle.

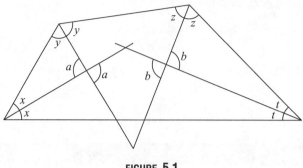

FIGURE **5.1**

5.2 Cyclic quadrilaterals with perpendicular diagonals

The next example shows how identifying angles can help one discover isosceles triangles, which then leads to the equality of certain segments.

Let $ABCD$ be a quadrilateral inscribed in a circle whose diagonals are perpendicular. Many such quadrilaterals exist, e.g., given a Cartesian coordinate system, all quadrilaterals obtained from the intersections of the x- and y-axes with a circle containing the origin will be in this class. Let P be the intersection of the diagonals. We'll now show that if a line going through P is perpendicular to one side of the quadrilateral, then the opposite side is bisected.

In a picture describing the given quadrilateral we mark complementary angles x and y at P and where these angles appear in the rest of the image:

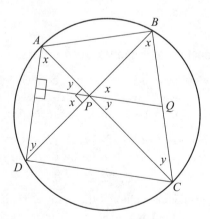

FIGURE **5.2**

Note that $\angle DAC$ and $\angle DBC$ both subtend arc DC, and $\angle ADB$ and $\angle ACB$ both subtend arc AB. Thus $\triangle PBQ$ is isosceles, so $\overline{BQ} = \overline{PQ}$, and $\triangle PQC$ is also isosceles, so $\overline{QC} = \overline{PQ}$, whence $\overline{BQ} = \overline{QC}$.

5.3 A property of the rectangular hyperbola

Here's a remarkable and little known property of the rectangular hyperbola.

Theorem. *Any line cutting one branch of the hyperbola $y = 1/x$ in two points A and B intersects the axes at points A' and B' such that the line segments AA' and BB' have the same length, i.e., $\overline{AA'} = \overline{BB'}$.*

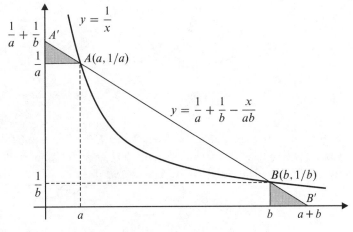

FIGURE **5.3**

The following proof, due to the first author and J. L. Garcia Roig, is a straightforward calculation of the lengths of relevant line segments.

Proof. The equation of the secant line through the points $A(a, 1/a)$ and $B(b, 1/b)$ is $y = 1/a + 1/b - x/ab$, which intercepts the axes at $(0, 1/a + 1/b)$ and $(a + b, 0)$. It now follows that the gray-shaded triangles are congruent, and hence their hypotenuses AA' and BB' have equal length.

Does the property hold for secant lines that cut both branches? See Challenge 5.4 below.

5.4 Challenges

5.1 Show that the vertex angles of a five-pointed star sum to 180° [Nakhli, 1986]. [Hint: draw the lines as shown in the second figure, and find equal angles.]

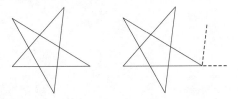

FIGURE **5.4**

5.2 The circle C_1 passes through the center O of the circle C_2, as shown below. Show that the length of the common chord PQ is equal to the length of the tangent segment PR [Eddy, 1992].

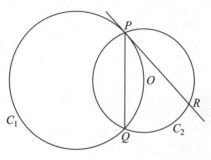

FIGURE **5.5**

5.2 Let A and B be the midpoints of the sides EF and ED of an equilateral triangle $\triangle DEF$. Extend AB to meet the circumcircle (of $\triangle DEF$) at C. Show that B divides AC according to the golden ratio [van de Craats, 1986].

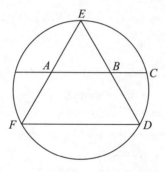

FIGURE **5.6**

5.4 Is the conclusion of the theorem in section 5.3 true if the secant line cuts both branches of the hyperbola?

6

Employing Isometry

Isometries in the plane are linear transformations that preserve distances. The basic isometries are rotations, translations, and reflections. Preservation of Euclidean distance yields the fact that isometries preserve angles and areas, so shapes of figures are invariant. Two geometric figures related by an isometry are congruent.

6.1 The *Chou pei suan ching* proof of the Pythagorean theorem

Perhaps the simplest (and most elegant) proof of the Pythagorean theorem is the following one from the *Chou pei suan ching*, a Chinese document dating to approximately 200 B.C. It uses only translations of triangles within a square:

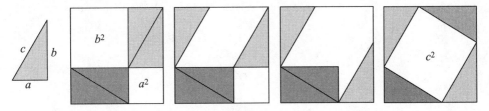

FIGURE 6.1

The total area of the white portions inside the large square remains unchanged as three of the four shaded triangles are translated to new positions.

6.2 A theorem of Thales

One of many geometric theorems attributed to Thales of Miletus (640–546 B.C.)—*corresponding sides of similar triangles are proportional*—is a keystone of the study of similarity of figures and the basis for defining trigonometric functions. Let's use isometry to prove this theorem for right triangles.

In Figure 6.2(a) we've superimposed the smaller of a pair of similar triangles on the larger, and we wish to establish that $a/a' = b/b'$. We first rotate the smaller triangle $180°$ about vertex O, and then draw a rectangle about the two triangles, as indicated by the dashed lines in Figure 6.2(b). The triangles above and below the diagonal of the rectangle are congruent and thus have the same area, from which it follows that the two gray rectangles have the same area, i.e., $a'b = ab'$. Hence $a/a' = b/b'$. Note: if c and c' denote the hypotenuses of the triangles, $c/c' = a/a' = b/b'$ follows by simple algebra.

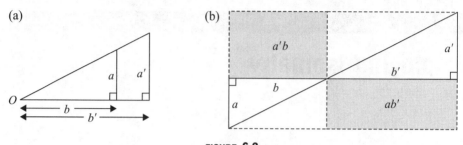

FIGURE 6.2

6.3   Leonardo da Vinci's proof of the Pythagorean theorem

In Figure 6.3 we have a visual proof of the Pythagorean theorem attributed to the remarkable Leonardo da Vinci (1542–1519). To the "standard" picture of the right triangle with squares on the legs and the hypotenuse, Leonardo strategically added two copies of the right triangle, and the two dashed lines CC' and DD'. By reflection isometry, $DEFD'$ is congruent to $DBAD'$, and by rotation isometry, $CBA'C'$ is congruent to $CAB'C'$. Observe that a $90°$ clockwise rotation of $DBAD'$ about point B shows that $DBAD'$ and $CBA'C'$ are also congruent. Hence the hexagons $DEFD'AB$ and $CAB'C'A'B$ have congruent parts and consequently the same area, from which the Pythagorean theorem follows.

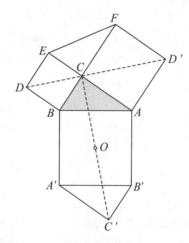

FIGURE 6.3

6.4   The Fermat point of a triangle

The *Fermat point* of an acute triangle is defined as follows: it is the point F inside $\triangle ABC$ such that sum $FA + FB + FC$ of distances from F to the vertices is a minimum (see Figure 6.4(a)). Rather surprisingly, the Fermat point can be located in the following manner: Construct equilateral triangles on the sides of $\triangle ABC$, as illustrated in Figure 6.4(b), and join each vertex of $\triangle ABC$ to the exterior vertex of the opposite equilateral triangle. Those three lines intersect at the Fermat point of $\triangle ABC$.

The following proof [Bogomolny, 1996] of that fact uses rotational isometry. In $\triangle ABC$, select any point P and connect P to the vertices A, B, and C. Now rotate $\triangle APB$ 60° counterclockwise to form $\triangle C'P'B$, and draw lines $C'A$ and $P'P$ [see Figure 6.4(c)].

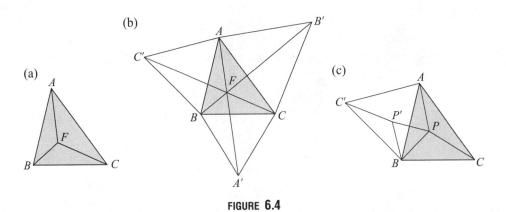

FIGURE **6.4**

Now $\triangle ABC'$ is equilateral, $PA = P'C'$, $PB = P'B$, and consequently $\triangle BPP'$ is equilateral, so that $PB = P'P$. Thus

$$PA + PB + PC = C'P' + P'P + PC.$$

This last sum will be a minimum when P' and P both lie on the line $C'C$ (note that, as the image of A under the rotation, the position of C' doesn't depend on P). Thus $PA + PB + PC$ is a minimum if and only if P lies on $C'C$, and for such a P, $\angle BPC' = 60°$. Since the choice of which side of the triangle to rotate was arbitrary, P must also lie on $B'B$ and $A'A$.

We will encounter another proof of this remarkable result in Chapter 19.

6.5   Viviani's theorem

The following theorem, attributed to Vincenzo Viviani (1622–1703), gives a remarkable property of equilateral triangles.

Theorem.   *The perpendiculars to the sides from a point on the boundary or within an equilateral triangle add up to the height of the triangle.*

The following proof [Kawasaki, 2005] uses only rotation to establish the theorem.

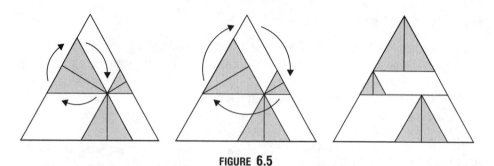

FIGURE **6.5**

6.6   Challenges

6.1 Use isometries to establish the *"Behold!"* proof of the Pythagorean theorem in Figure 6.6, attributed to the 12th century Hindu mathematician Bhāskara.

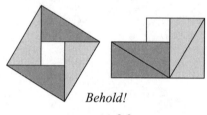

Behold!

FIGURE **6.6**

6.2 Using Figure 6.7, give another proof of Viviani's theorem using reflection rather then rotation.

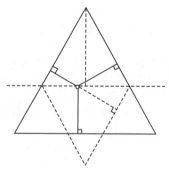

FIGURE **6.7**

6.3 Find the area of a convex octagon that is inscribed in a circle and has four consecutive sides of length 3 and the remaining four sides of length 2 units.

6.4 If F is the Fermat point of $\triangle ABC$ (see Figure 6.4(b)), show that $\angle AFB = \angle BFC = \angle CFA = 120°$.

7

Employing Similarity

In this chapter we explore applications of similarity of geometric figures—primarily triangles—as a tool for illustrating and proving theorems in geometry. We've employed this technique in earlier demonstrations, for example, in Section 2.1, where we used similar triangles in Figure 2.2(a) to establish the fact that the altitude to the hypotenuse in a right triangle is the geometric mean of the segments it determines on the hypotenuse.

7.1 Ptolemy's theorem

The following theorem—and proof, using only similar triangles—is due to Ptolemy of Alexandria (circa 150 A.D.).

Theorem. *In a quadrilateral inscribed in a circle, the product of the lengths of the diagonals is equal to the sum of the products of the lengths of the opposite sides.*

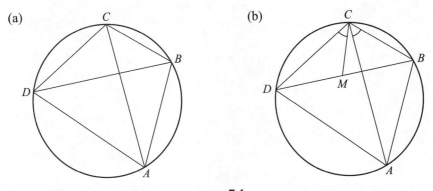

FIGURE 7.1

Figure 7.1(a) illustrates a general quadrilateral inscribed in a circle, and in Figure 7.1(b) we draw the line segment CM so that $\angle DCM$ equals $\angle ACB$. Angles $\angle CDB$

and $\angle CAB$ subtend the same arc CB, hence they are equal, and thus triangles $\triangle DCM$ and $\triangle ACB$ are similar (see Figure 7.2(a)). Consequently $\overline{CD}/\overline{MD} = \overline{AC}/\overline{AB}$, so that $\overline{AB} \cdot \overline{CD} = \overline{AC} \cdot \overline{MD}$.

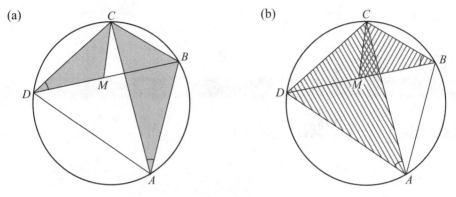

FIGURE 7.2

Similarly, angles $\angle DAC$ and $\angle DBC$ subtend the same arc DC, hence they are equal, and thus triangles $\triangle DAC$ and $\triangle CBM$ are similar (see Figure 7.2(b)). Thus $\overline{BC}/\overline{BM} = \overline{AC}/\overline{AD}$, so that $\overline{BC} \cdot \overline{AD} = \overline{AC} \cdot \overline{BM}$. Adding, we have $\overline{AB} \cdot \overline{CD} + \overline{BC} \cdot \overline{AD} = \overline{AC}(\overline{MD} + \overline{BM}) = \overline{AC} \cdot \overline{BD}$.

7.2   The golden ratio in the regular pentagon

The golden ratio—$\phi = (1 + \sqrt{5})/2 \simeq 1.618$—appears in many surprising contexts in geometry. One such occurrence is the following: It is the length x of a diagonal in a regular pentagon whose sides have length 1, as illustrated in Figure 7.3(a).

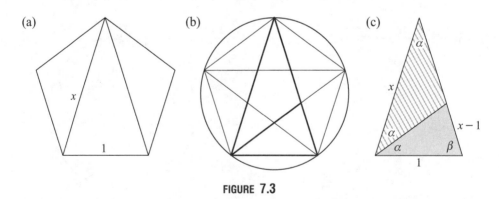

FIGURE 7.3

Two such diagonals from a common vertex form an isosceles triangle, which can be partitioned by a segment of another diagonal into two smaller triangles, as illustrated in Figure 7.3(b) and (c). The angles labeled α in Figure 7.3(c) are equal since each subtends an arc equal to 1/5 the circumference of the circumscribing circle. Thus the striped triangle is also isosceles, $\beta = 2\alpha$, and so $\alpha = 36°$, since the sum of the angles in the original triangle yields $5\alpha = 180°$. It now follows that the unlabeled angle in the gray triangle is $\beta = 72°$, and hence it is also isosceles and similar to the original triangle. Hence $x/1 = 1/(x-1)$, and so x is the (positive) root of $x^2 - x - 1 = 0$, that is, $x = (1 + \sqrt{5})/2$ as claimed.

7.3 The Pythagorean theorem—again

As noted in Section 2.1, the altitude to the hypotenuse of a right triangle partitions the triangle into two smaller triangles, each similar to the original. This provides the following proof of the Pythagorean theorem:

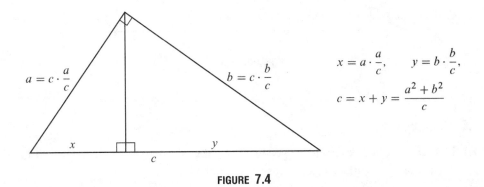

$$x = a \cdot \frac{a}{c}, \qquad y = b \cdot \frac{b}{c},$$

$$c = x + y = \frac{a^2 + b^2}{c}$$

FIGURE 7.4

7.4 Area between sides and cevians of a triangle

A *cevian* is a line segment joining a vertex of a triangle with a point on the opposite side (or its extension). In an equilateral triangle, draw two cevians from the base vertices to points 1/3 the way to the top vertex, as shown in Figure 7.5. What is the relationship between the areas of the original triangle and the shaded triangle bounded by the cevians and the base?

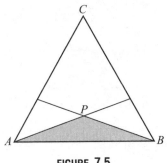

FIGURE 7.5

Let $[ABC]$ denote the area of a triangle $\triangle ABC$. In Figure 7.6, we use five copies of $\triangle ABC$, and observe that $\triangle APQ$ is similar to $\triangle ADE$, and that $[APQ] = (1/25)[ADE]$. Furthermore, $[ADE] = (5/2)[ABC]$. Hence

$$[APQ] = (1/25)[ADE] = (1/25)(5/2)[ABC] = (1/10)[ABC],$$

and thus $[APB] = (1/5)[ABC]$.

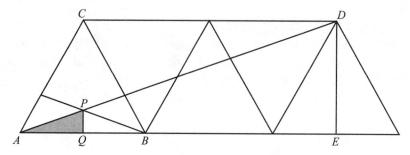

FIGURE **7.6**

7.5   Challenges

7.1 If K, a, b, c, and R denote, respectively, the area, the lengths of the sides, and the circumradius of a triangle, show that $K = abc/4R$.

7.2 Show that $\sin 54° = \cos 36° = \phi/2$ and $\sin 18° = \cos 72° = (\phi - 1)/2$. [Hint: Start with Figure 7.3(c).]

7.3 Prove the **Reciprocal Pythagorean Theorem:** *If a and b are the legs and h the altitude to the hypotenuse of a right triangle, then*

$$\frac{1}{a^2} + \frac{1}{b^2} = \frac{1}{h^2}.$$

7.4 Connect the midpoints and the "1/3-points" of the sides of an equilateral triangle, as shown in Figure 7.7, to create an interior hexagon (shaded). Show that the area of the hexagon is 2/5 the area of the triangle.

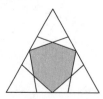

FIGURE **7.7**

8

Area-preserving Transformations

The basic length-preserving transformations in the plane are the isometries, which we examined in Chapter 6. Since isometries preserve lengths, they necessarily preserve angles, areas, volumes, etc. We now consider transformations in the plane which may not preserve lengths and angles, but do preserve areas.

First consider triangles and parallelograms. If two triangles have a common base and if their vertices lie on a line parallel to the base, they must have equal areas, as illustrated in Figure 8.1(a). Similarly, two parallelograms with a common base and equal heights also

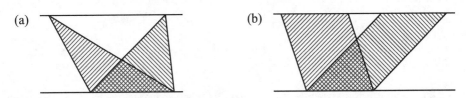

(a)   (b)

FIGURE **8.1**

have the same area, as illustrated in Figure 8.1(b). We use these simple transformations to prove two important theorems in the next section.

8.1 Pappus and Pythagoras

In Book IV of his *Mathematical Collection*, Pappus of Alexandria (circa 320 A.D.) stated the following generalization of the Pythagorean theorem.

Theorem. *Let ABC be any triangle, and $ABDE$, $ACFG$ any parallelograms described externally on AB and AC. Extend DE and FG to meet in H, and draw BL and CM equal and parallel to HA. Then, in area, $BCML = ABDE + ACFG$. See Figure 8.2.*

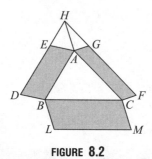

FIGURE 8.2

The proof below uses successive area-preserving transformations of the parallelograms:

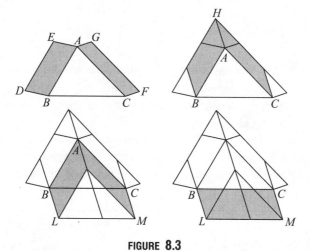

FIGURE 8.3

If the triangle is a right triangle, and the parallelograms squares, we have the following proof of the Pythagorean theorem:

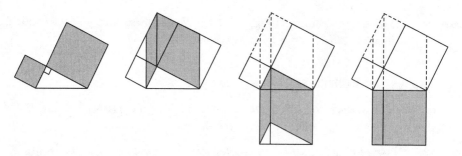

FIGURE 8.4

8.2 Squaring polygons

One of the classical problems of Greek geometry concerned "squaring" a figure: given a figure, construct a square with the same area in a finite number of steps using only a compass and an unmarked straightedge. As we know, it is impossible to square a circle, but we will show how to square any convex n-gon.

There are several visual approaches to squaring convex n-gons. The first is inductive, so we begin with triangles.

Lemma. *Any triangle can be squared with compass and straightedge.*

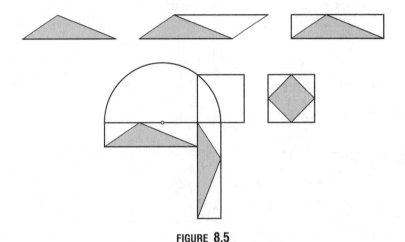

FIGURE **8.5**

One passes from the triangle to the parallelogram formed by two copies of the triangle; then to the rectangle of equal area; then from the rectangle to a square whose side is the geometric mean of the dimensions of the rectangle, and hence has the same area (see Figure 2.2(a)); and finally to a square with half the area of the first.

Theorem. *Given a convex n-gon P_n, $n \geq 4$, one can construct with compass and straightedge a convex $(n-1)$-gon P_{n-1} with the same area as P_n.*

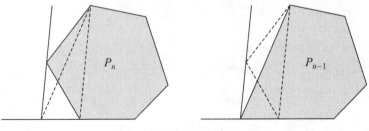

FIGURE **8.6**

The construction in Figure 8.6 is self-explanatory, except to note that the line through the left-most vertex of P_n is drawn parallel to the dashed diagonal.

Another possibility for squaring n-gons is also quite simple: divide the n-gon into triangles using diagonals from one vertex, and square each triangle using the lemma above. Then add the squares, using the Pythagorean theorem, to form a single square.

8.3 Equal areas in a partition of a parallelogram

Take a parallelogram, and from arbitrary points on two adjacent sides, draw lines to the vertices on the opposite sides, as shown in Figure 8.7(a). This partitions the parallelogram into eight regions with areas $a, b, \ldots, h$ as indicated. The following sequence of area-preserving transformations (Figure 8.7(b)) shows that $a + b + c = d$ [Richard, 2004]:

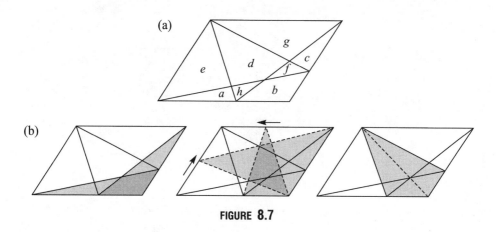

FIGURE **8.7**

Equating areas of triangles yields $a + h + 2b + f + c = b + h + f + d$, which is equivalent to the desired result. In Challenge 8.6, you will show that $e + f = g + h$.

8.4 The Cauchy-Schwarz inequality

For any real numbers a, b, x, y, the Cauchy-Schwarz inequality states that

$$|ax + by| \leq \sqrt{a^2 + b^2} \cdot \sqrt{x^2 + y^2}.$$

Using properties of absolute value, we have $|ax + by| \leq |a||x| + |b||y|$, so to establish the Cauchy-Schwarz inequality, it suffices to show that

$$|a||x| + |b||y| \leq \sqrt{a^2 + b^2} \cdot \sqrt{x^2 + y^2}.$$

In Figure 8.8, we again use the result of Pappus from Section 8.1, starting with a right triangle whose legs have lengths $|a|$ and $|b|$, and two rectangles of dimensions $|a|$ by $|x|$ and $|b|$ by $|y|$. We obtain a parallelogram whose sides have lengths $\sqrt{a^2 + b^2}$ and $\sqrt{x^2 + y^2}$, and with area less than or equal to that of the rectangle with sides of the same lengths.

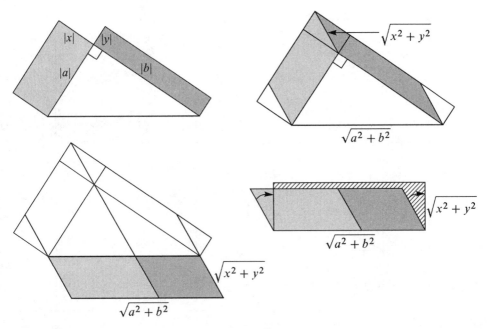

FIGURE **8.8**

8.5 A theorem of Gaspard Monge

The French geometer Gaspard Monge (1746–1818) noticed a curious property concerning some triangles associated with a line of positive slope in the first quadrant. To be precise, let AB denote a segment of the line, and M its midpoint. Construct triangles with vertex M as shown in Figure 8.9(a), and denote their areas by S_x and S_y. Also construct $\triangle OAB$, as shown in Figure 8.9(b), and let S denote its area Then $S = |S_x - S_y|$.

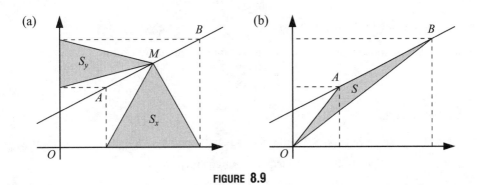

FIGURE **8.9**

[Note: When the y-intercept of the line is positive, $|S_x - S_y| = S_x - S_y$; and when the x-intercept is positive, $|S_x - S_y| = S_y - S_x$. We consider the first case, the second is similar.] The sequence of area-preserving transformations in Figure 8.10 proves the

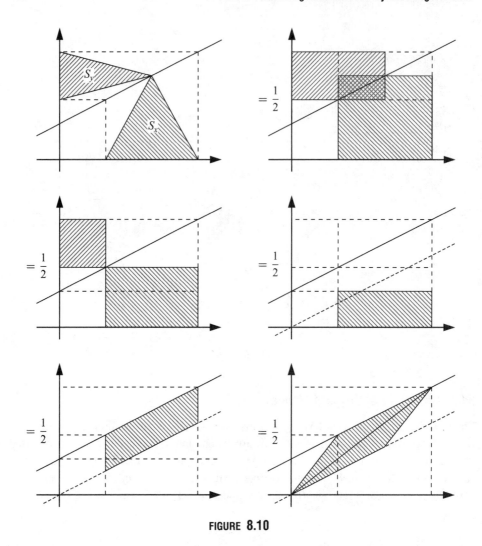

FIGURE **8.10**

theorem. The different shading for S_x and S_y indicates that S_y will be subtracted from S_x.

8.6   Challenges

8.1 Can the ideas in Section 8.2 be used to square non-convex n-gons?

8.2 When is the Cauchy-Schwarz inequality an equality?

8.3 Can the circle be squared with drawing tools different from compass and straightedge?

8.4 Use an area-preserving transformation to establish the following formula for factoring the sum of two squares:

$$x^2 + y^2 = \left(x + \sqrt{2xy} + y\right)\left(x - \sqrt{2xy} + y\right).$$

8.5 Use Figure 8.11 and the observation that the shaded parallelograms have equal area to derive the sine law, i.e.,

$$\frac{\sin \alpha}{a} = \frac{\sin \beta}{b}.$$

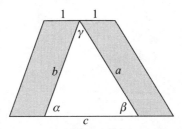

FIGURE 8.11

8.6 Use an area-preserving transformation to establish $e + f = g + h$ for the partitioned parallelogram in Section 8.3.

8.7 Given an isosceles triangle as shown in Figure 8.12, show that $c^2 = a^2 + bd$.

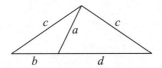

FIGURE 8.12

9

Escaping from the Plane

We have a natural tendency to solve planar problems in the plane and spatial questions in space. But restricting our arguments to the given context (plane or space) may be too limiting! There are many examples of planar problems which are easier to solve if we look at them "from a spatial point of view." Conversely, one can also find spatial problems that may be more easily solved when reduced to planar problems.

Our message in this chapter is to keep in mind the possibility of combining planar and spatial techniques.

9.1 Three circles and six tangents

Let's consider a beautiful result in the plane, known as *Monge's Circle Theorem* [Bogomolny, 1996]: given mutually disjoint circles of different diameters, consider the tangent lines to each pair of circles. These pairs of tangents determine three intersection points. The surprising result is that these three points lie on a line L. The following drawing illustrates the claim.

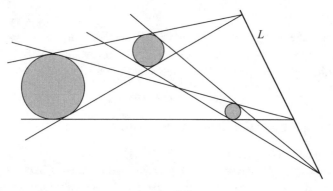

FIGURE 9.1

It is not so easy to prove this statement in Flatland. The following spatial argument turns the question into a trivial situation in three-dimensional space.

Let's look at Figure 9.1 from space: let each circle be the "equator" of a sphere. Given a pair of spheres consider the cone generated by the two corresponding tangent lines. Half of the cone will lie above the plane of the circles and half will lie below. Now consider a plane tangent to the three half-spheres. This plane will also be tangent to each of the three cones, and it will intersect the original plane in a line L. Since this plane contains one line from each half-cone, the vertices of the three cones must be located on the intersection line L.

9.2   Fair division of a cake

Suppose we have a rectangular cake, with chocolate icing on top and all the sides, and we would like to share this cake among five people. Our aim is to find a fair division of the cake where "fair" means that each person receives the some volume of cake and, of course, the same amount of chocolate icing. The following solution [Sanford, 2002] is a clever one:

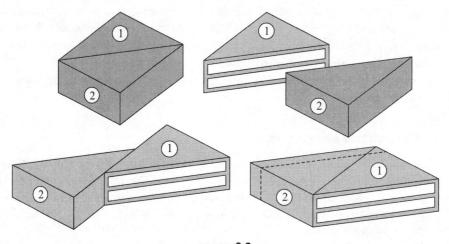

FIGURE **9.2**

You cut the cake along a diagonal, move half cake as indicated (a spatial move) and then cut the top parallelogram into five parts (a planar division, the first cut is given by the dashed line). Each person gets two pieces and the total amount of cake and icing is the same—a fair division!

9.3   Inscribing the regular heptagon in a circle

One of the famous classical problems of geometry is the impossibility of inscribing the regular heptagon in a given circle by means of straightedge and compass. We will illustrate here an exact solution working in space with two additional tools—scissors and cardboard!

Given a circle of radius R, draw on cardboard a circle of radius $8R/7$ divided into 8 parts. Cut out this circle with scissors and throw away one slice. Then form the remaining piece into a cone. Since the circumference of the base of the cone is the same as that of the original circle, the cone fits exactly on the top of the circle and divides it into 7 equal parts, thus inscribing a regular heptagon in the circle!

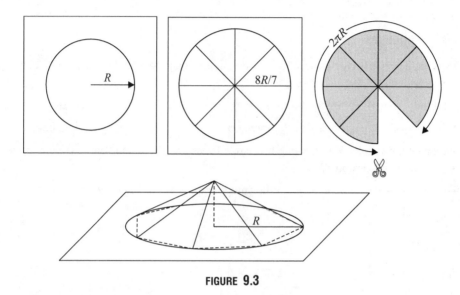

FIGURE **9.3**

9.4   The spider and the fly

Here's a three-dimensional problem that's easier to solve in two dimensions. It was created by Henry Ernest Dudeney (1857–1931), and is perhaps his best-known brain teaser [Gardner, 1961]. A spider and a fly are in a room 30 feet long, 12 feet high and 12 feet wide, as illustrated in Figure 9.4(a). The spider is in the middle of an end wall, one foot from the floor (the black dot); while the fly is in the middle of the opposite end wall, one foot from the ceiling (the gray dot), too paralyzed with fear to move. What is the shortest distance the spider must crawl to reach the fly?

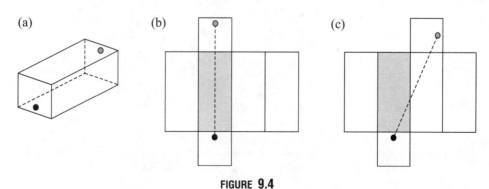

FIGURE **9.4**

To find the shortest path, we "unfold" the room to get a plane diagram on which to trace the path between the spots. Two of several feasible ways to unfold the room are shown in Figures 9.4(b) and (c) (the gray rectangle is the floor of the room). The distance from the spider to the fly in (b) is 42 feet, while the distance in (c) is only (using the Pythagorean theorem) $\sqrt{1658} \simeq 40.7$ feet. We leave it as a challenge to determine if this path is the shortest. [Hint: it's not!]

9.5 Challenges

9.1 Find alternative methods for the fair division among five people of the cake described in Section 9.2.

9.2 Find a spatial representation for $(a+b)^3 = a^3 + 3a^2b + 3ab^2 + b^3$ for any $a, b \geq 0$.

9.3 Show how to visualize that the intersection of two spheres in space is a circle.

9.4 Show how to inscribe an n-gon in a circle for any $n \geq 3$ (using straightedge, compass, scissors and cardboard).

9.5 Find the shortest path between the spider and the fly in Figure 9.4(a).

10

Overlaying Tiles

A *tiling* of the plane is a countable family of closed sets (the *tiles*) that cover the plane without gaps or overlaps [Grünbaum and Shepard, 1986]. In Figure 10.1 we see portions of two examples, composed of tiles that are squares of two different sizes in (a) and rectangles and squares in (b).

(a)   (b)

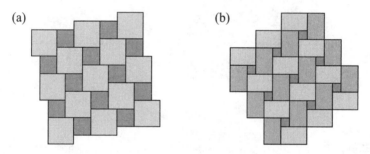

FIGURE **10.1**

Indeed, tilings such as those above have been used for many centuries in homes, churches, palaces, etc. [Eves, 1976]. If we overlay a second grid of "transparent" tiles, we can construct visual demonstrations of a variety of mathematical theorems. We begin with the tilings in Figure 10.1, which yield several classic proofs of the Pythagorean theorem.

10.1   Pythagorean tilings

We begin with the tiling in Figure 10.1(a), and overlay a grid of transparent square tiles, as shown in Figure 10.2(a). Note that the side of the square in the overlay is the hypotenuse of a right triangle whose legs are the sides of the two smaller squares in the original tiling. As seen in Figure 10.2(b), the tiling and overlay generate a "dissection" proof of the Pythagorean theorem, wherein one sees how the squares on the legs of the triangle should be dissected and reassembled to form the square on the hypotenuse. This proof is usually attributed to Annairizi of Arabia (circa 900 A.D.) [Annairizi].

47

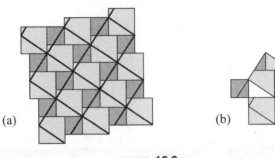

(a)   (b)

FIGURE **10.2**

If we translate the overlay grid of squares so that the vertices of the overlay squares coincide with the centers of the larger squares in the original tiling, we obtain a second dissection proof of the Pythagorean theorem. This proof, in Figure 10.3, is often attributed to Henri Perigal (1801–1899). Any other translated position for the overlay grid yields another proof—indeed, there are uncountably many different such dissections proofs of the Pythagorean theorem constructed from the tiling in Figure 10.1(a)!

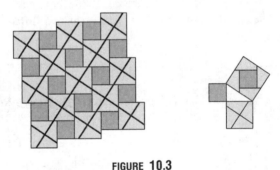

FIGURE **10.3**

The tiling in Figure 10.1(b), with an overlay of square tiles whose sides are diagonals of the rectangles in the original tiling, is the basis for the well-known "Behold!" proof of the Pythagorean theorem attributed to Bhāskara (12th century A.D.), as illustrated in Figure 10.4 [Eves, 1980]. In Challenge 6.1 we used translation isometry to establish the same result.

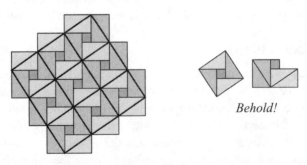

Behold!

FIGURE **10.4**

If we tile the plane in the same general pattern as illustrated in Figure 10.1(b), using three different rectangles two of which have diagonals of length 1, and overlay a grid of rhombi, we obtain a proof of the "sine of the sum" trigonometric identity [Priebe and Ramos, 2000].

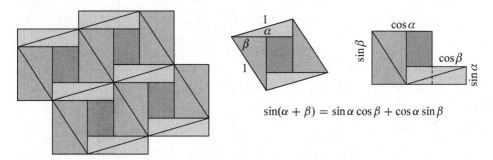

$$\sin(\alpha + \beta) = \sin\alpha\cos\beta + \cos\alpha\sin\beta$$

FIGURE **10.5**

10.2 Cartesian tilings

Tiling the plane with squares all the same size resembles ordinary graph paper, so we call it a Cartesian tiling (in general, when all the tiles in a tiling have the same size and shape, it is called a *monohedral* tiling). If we overlay the same grid of transparent square tiles used in Figures 10.2, 10.3, and 10.4, we obtain a proof of the following theorem, as seen in Figure 10.6.

Theorem.   *If lines from the vertices of a square are drawn to the midpoints of adjacent sides, then the area of the smaller square so produced is one-fifth that of the given square.*

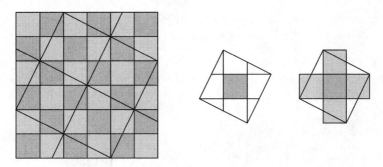

FIGURE **10.6**

With a different overlay (not a tiling), the same Cartesian tiling yields a proof (in Figure 10.7) of the following theorem.

Theorem.   *A square inscribed in a semicircle has two-fifths the area of a square inscribed in a circle of the same radius.*

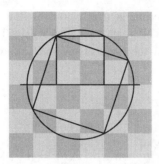

FIGURE **10.7**

10.3   Quadrilateral tilings

So far, the tilings we've considered have used squares and rectangles. However, copies
of *any* quadrilateral, concave or convex, will also yield a monohedral tiling of the plane,
as illustrated in Figure 10.8.

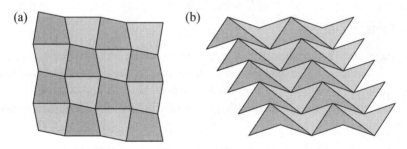

FIGURE **10.8**

With an overlay of transparent parallelogram tiles, the tiling in Figure 10.8(a) is used
in Figure 10.9 to prove the following theorem.

Theorem.   *The area of any convex quadrilateral Q is equal to one-half the area of a
parallelogram P whose sides are parallel to and equal in length to the diagonals of Q.*

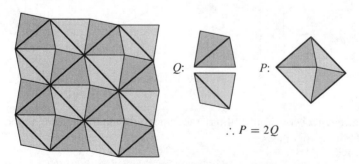

FIGURE **10.9**

This theorem provides perhaps the easiest way to compute the area of a general quadrilateral! In fact, the theorem is also true for concave quadrilaterals even though one of the diagonals is exterior to the quadrilateral—see Challenge 10.2.

10.4 Triangular tilings

Just as with quadrilaterals, copies of an arbitrary triangle will also form a monohedral tiling of the plane. The following theorem is a "triangular" analog of the first theorem in Section 10.2, and its proof is illustrated in Figure 10.10 with a triangular tiling and a triangular overlay [Johnston and Kennedy, 1993].

Theorem. *If the one-third points on each side of a triangle are joined to opposite vertices, the resulting triangle is equal in area to one-seventh that of the original triangle.*

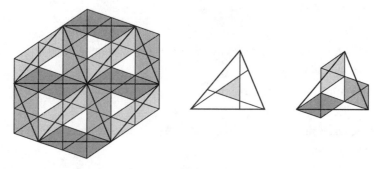

FIGURE **10.10**

10.5 Tiling with squares and parallelograms

Take an arbitrary parallelogram and construct squares externally on each of the sides. What sort of quadrilateral is formed by joining the centers of the squares? To answer the question, extend the construction to a tiling, as shown in Figure 10.11 [Flores, 1997].

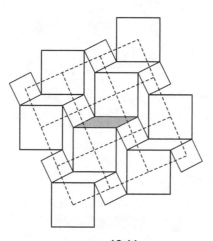

FIGURE **10.11**

10.6 Challenges

10.1 Suppose we draw lines from the vertices of a square to points one-third the way from an adjacent vertex to the opposite vertex, as shown in Figure 10.12. Use a Cartesian tiling to determine the area of the smaller square so formed. What if the point is two-thirds the way from the adjacent vertex? What about other fractions?

FIGURE **10.12**

10.2 Use the tiling in Figure 10.8(b) to prove that the theorem in Section 10.3 holds for concave quadrilaterals.

10.3 Use the tiling in Figure 10.5 to prove the "cosine of the difference" formula: $\cos(\alpha - \beta) = \cos\alpha\cos\beta + \sin\alpha\sin\beta$.

10.4 Can the result in the theorem in Section 10.4 be generalized in a manner similar to that in Challenge 10.1?

10.5 One equilateral triangle is inscribed in another, as shown in Figure 10.13. Use an overlay tiling of equilateral triangles to show that the area of the smaller triangle is one-third that of the larger triangle.

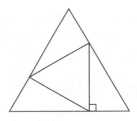

FIGURE **10.13**

10.6 Use a triangular monohedral tiling and overlay to show that the medians of a triangle form a new triangle with three-fourths the area of the original triangle, as illustrated below.

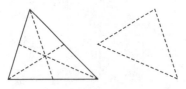

FIGURE **10.14**

10.7 Below is the tiling that corresponds to the "triangular" version of the result in Section 10.5. From the tiling, can you state the corresponding result about triangles? It is known as *Napoleon's theorem,* although it is doubtful that Napoleon knew enough geometry to either state or prove the theorem [Coxeter and Greitzer, 1967; Coxeter, 1969].

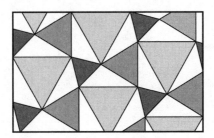

FIGURE **10.15**

11

Playing with Several Copies

In Chapters 3 and 4 we saw several instances where making several copies of the figure representing a mathematical expression made it easier to evaluate that expression. For example, we used two copies of an area representation of the sum $1 + 2 + \cdots + n$ in Figure 3.2 to show that the sum was one-half the area of a rectangle, and we used two copies of a volume representation for the double sum $\sum_{i=1}^{n} \sum_{j=1}^{n} (i + j - 1)$ in Figure 4.6 to show that the value of the double sum was one-half the volume of a rectangular box.

In this chapter we present additional examples of this technique, using more than two copies. We begin with a result in trigonometry based on a proof of the Pythagorean theorem in a previous section.

11.1 From Pythagoras to trigonometry

In Section 6.1 we saw the elegant *Chou pei suan ching* proof of the Pythagorean theorem, which employed four copies of a right triangle. That proof can be easily modified, using multiple copies of two different triangles, to give a proof of the addition formula for the sine. In this proof [Priebe and Ramos, 2000], we appeal once again to the Fubini principle, and compute the areas of the shaded regions.

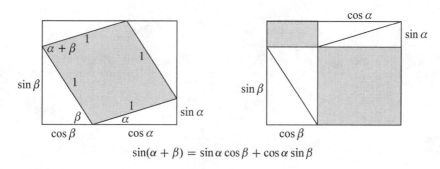

$$\sin(\alpha + \beta) = \sin\alpha\cos\beta + \cos\alpha\sin\beta$$

FIGURE 11.1

11.2   Sums of odd integers revisited

As we saw in Section 1.1 (and Challenge 3.1), the sum of the first n odd numbers is n^2. If we represent $1 + 3 + \cdots + (2n - 1)$ by a "triangle" of unit squares, as shown below in Figure 11.2(a), then four copies of that "triangle" form a square with side length $2n$.

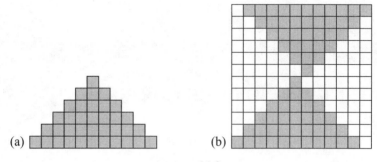

(a)   (b)

FIGURE 11.2

Thus $1 + 3 + \cdots + (2n - 1) = (2n)^2/4 = n^2$.

11.3   Sums of squares again

In Chapter 3 we saw (Figure 3.8) an area representation of the formula for the sum of the squares of the first n integers: $1^2 + 2^2 + \cdots + n^2 = n(n + 1)(2n + 1)/6$, in which three copies of $1^2 + 2^2 + \cdots + n^2$ were rearranged to form a rectangle. In that demonstration we needed to dissect one set of squares in order to reassemble the pieces into a rectangle. Since the sum $n(n + 1)(2n + 1)/6$ is cubic in n, perhaps a volume representation is in order. The denominator suggests using multiple copies, which leads to the following proof that $1^2 + 2^2 + \cdots + n^2 = n(n + \frac{1}{2})(n + 1)/3$ [Siu, 1984].

In Figure 11.3 we see how three copies of $1^2 + 2^2 + \cdots + n^2$ can be oriented to form a solid with a rectangular base, and when the cubes in the top layer are halved and moved, the result is a rectangular box with dimensions n by $n + 1$ by $n + (1/2)$.

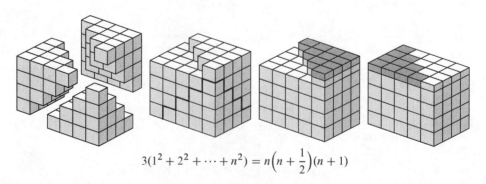

$$3(1^2 + 2^2 + \cdots + n^2) = n\left(n + \frac{1}{2}\right)(n + 1)$$

FIGURE 11.3

11.4   The volume of a square pyramid

Virtually the same idea presented above to sum squares can be employed to find the volume of a square pyramid. Let P_0 be a pyramid with a square base of side length b and height h, as illustrated in Figure 11.4. By Cavalieri's Principle, P_0 has the same volume as pyramid P_1, in which two of the triangular sides are perpendicular to the base. Pyramid P_2 has the same property as pyramid P_1, however its height is also the base side length b. Finally, three copies of P_2 fit together to form a cube (since a cube of side length b can be readily dissected to form three copies of P_2).

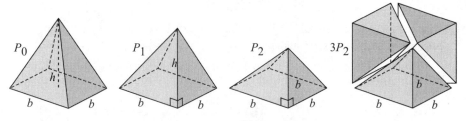

FIGURE **11.4**

Hence we have $V(P_0) = V(P_1) = \dfrac{h}{b} V(P_2) = \dfrac{h}{b} \cdot \dfrac{1}{3} b^3 = \dfrac{1}{3} b^2 h.$

11.5   Challenges

11.1 Prove that the internal bisector of the right angle of a right triangle bisects the square on the hypotenuse [Eddy, 1991].

FIGURE **11.5**

11.2 Use multiple copies Figure 11.6 to show that the sum of the cubes of the first n integers is $\frac{1}{4}\left[n(n+1)\right]^2$. [Cupillari, 1989; Lushbaugh, 1965].

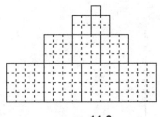

FIGURE **11.6**

11.3 Modify the method of Section 11.4 to show that the volume of a frustum of a square pyramid with bases of side length a and b, and height h, is $h(a^2 + ab + b^2)/3$. [Hint: $a^2 + ab + b^2 = (b^3 - a^3)/(b - a)$.]

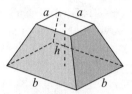

FIGURE **11.7**

11.4 Use the picture in Figure 11.1 (but with different labels) to prove the "cosine of the difference" formula (see Challenge 10.4).

12

Sequential Frames

In many instances a sequence of pictures can be employed to illustrate a particular idea in mathematics. We used this idea in several sections in Chapters 4, 8, and 11, for example. One can think of a sequence of pictures as stills or "frames" from a motion picture, or for a Java-driven illustration on the web. In this chapter we provide a few more examples of this technique.

12.1 The parallelogram law

Did you know that in any parallelogram, the sum of the squares on the diagonals is equal to the sum of the squares on the sides? See Figure 12.1:

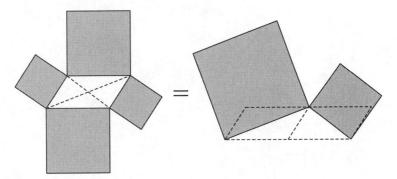

FIGURE **12.1**

In the next four figures we show with a sequence of frames how the four squares on the sides can be transformed, using isometries and the Pythagorean theorem, into a square on each of the diagonals.

Note that in moving from Figure 12.3 to Figure 12.4, we moved two lighter gray rectangles and a darker gray square from the left side of the figure to the right.

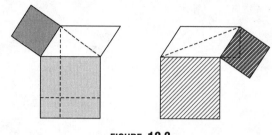

FIGURE **12.2**

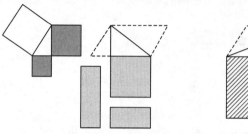

FIGURE **12.3**

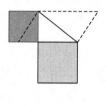
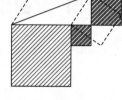
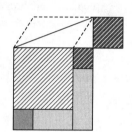

FIGURE **12.4**

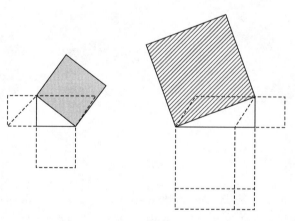

FIGURE **12.5**

12.2 An unknown angle

Consider the following problem: A point P is located within a square so that its distance from three consecutive vertices of the square is 1, 2, and 3 units, as illustrated below. What is the degree measure of the angle at P between the segments of lengths 1 and 2?

FIGURE **12.6**

You may wish try to find your own solution before reading further. We present a clever solution, devised by Murray Klamkin, which begins by making a copy of the square and rotating it 90° counterclockwise, and then placing it on the left side of the original square, as illustrated in Figure 12.7.

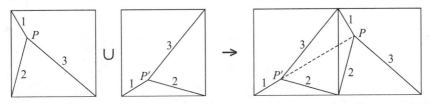

FIGURE **12.7**

Since the two segments of length 2 are perpendicular, they form the legs of an isosceles right triangle, whose hypotenuse $P'P$ has length $2\sqrt{2}$. But now the segments of lengths 1, $2\sqrt{2}$, and 3 also form a right triangle, and hence the angle at P between the segments of lengths 1 and 2 is $45° + 90° = 135°$.

12.3 Determinants

If two vectors are drawn from the origin to points (a, b) and (c, d) in the plane, then the area of the parallelogram generated by these vectors is the absolute value of a 2×2 determinant. See Figure 12.8.

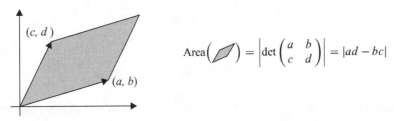

$$\text{Area}\left(\parallelogram\right) = \left|\det\begin{pmatrix} a & b \\ c & d \end{pmatrix}\right| = |ad - bc|$$

FIGURE **12.8**

We prove the result in Figure 12.9 (for the case where the parallelogram is in the first quadrant) with a pair of frames, again employing isometries and denoting subtraction of an area by striped shading [Golomb, 1965].

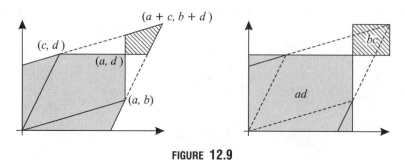

FIGURE 12.9

In the frame on the left, two triangular regions whose longest sides are the segments joining (a, b) to $(a+c, b+d)$ and (c, d) to $(a+c, b+d)$ have been shifted to positions adjacent to the axes. In the frame on the right, smaller triangles have been added and subtracted to complete the rectangles.

12.4 Challenges

12.1 Given any triangle in the plane, construct squares externally on each side, and connect vertices of adjacent squares to form three new triangles, as illustrated in Figure 12.10. Show that each of the new triangles has the same area as the original triangle. [Hint: rotate each new triangle counterclockwise about the vertex it shares with the original triangle.]

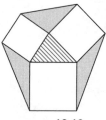

FIGURE 12.10

12.2 A point Q is located within a square so that its distance from three consecutive vertices of the square is 1, $\sqrt{3}$, and $\sqrt{5}$ units, as illustrated in Figure 12.11. What is the degree measure of the angle at Q between the segments of lengths 1 and $\sqrt{3}$?

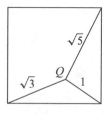

FIGURE 12.11

12.3 Show that the result in Section 12.3 holds no matter in which quadrant the terminal points (a, b) and (c, d) of the vectors lie.

13

Geometric Dissections

Geometric dissections have long been popular in recreational mathematics. The basic task in these recreational puzzles is to cut up a geometric figure into pieces and reassemble the pieces to form a different figure. Sam Loyd (1841–1911) created a great many dissection puzzles, such as the "sedan chair" in Figure 13.1, where one was asked to "cut the sedan chair into the fewest possible pieces, which will fit together and form a perfect square, so the men will appear to be carrying a closed box" (see Challenge 13.1 at the end of this chapter). The "tangram" puzzles in Part II are based on geometric dissections.

The process of dissecting and reassembling is another way to produce visual displays based on the principle of area-preserving transformations. For example, Figures 10.2(b) and 10.3(b) illustrate "dissection proofs" of the Pythagorean theorem, as do the two below

FIGURE **13.1**

in Figure 13.2, created by J. E. Böttcher and Liu Hui, respectively [Nelsen, 2000b], wherein the squares on the legs of the triangle are dissected and reassembled to form the square on the hypotenuse.

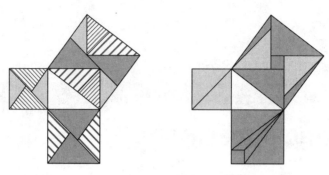

FIGURE **13.2**

13.1   Cutting with ingenuity

Here's a simple dissection problem: given an "L" shaped region formed from three-quarters of a square, show how it can be dissected into 2, 3, 4, 8, and 12 congruent pieces. A set of solutions is shown in Figure 13.3.

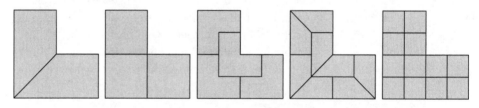

FIGURE **13.3**

A natural question is: For which values of n (besides 2, 3, 4, 8, and 12) can the "L" be dissected into n congruent pieces? This question is explored in Challenge 13.3. A word of caution: irregular figures or figures with holes can be difficult to dissect in prescribed ways. For example, suppose we require a dissection of the figure on the left in Figure 13.4 into two congruent pieces. A solution is given on the right.

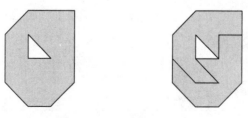

FIGURE **13.4**

13.2   The "Smart Alec" puzzle

This is another famous puzzle due to Sam Loyd. The figure is a concave pentagon formed
by removing an isosceles right triangle from a square, as shown in Figure 13.5.

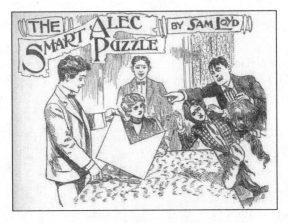

FIGURE 13.5

The task proposed by Loyd was to dissect the figure into four parts which could be
reassembled to form a square. Loyd's solution is given in Figure 13.6.

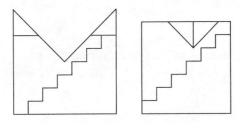

FIGURE 13.6

However, Loyd's solution is faulty—a simple calculation shows that the "square" on
the right is actually a rectangle. Indeed, no four-piece solution to the puzzle is known.
However, H. E. Dudeney found the beautiful five-piece solution given in Figure 13.7.

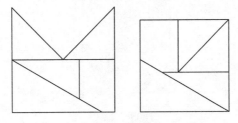

FIGURE 13.7

13.3   The area of a regular dodecagon.

Since the area of the unit circle is π and it's circumference is 2π, the area of any regular polygon inscribed in the unit circle will be less than π, and its perimeter will be less than 2π. By inscribing and circumscribing polygons with 6, 12, 24, 48, and 96 sides, Archimedes was able to show that π lies between $3\frac{10}{71}$ and $3\frac{1}{7}$. Archimedes used the fact that the area of the inscribed regular $2n$-gon is numerically the same as half the perimeter of the inscribed regular n-gon. For example, the perimeter of the regular hexagon (the "6-gon") inscribed in the unit circle is 6, and the area of the inscribed regular dodecagon (the "12-gon") is 3. Here is a clever dissection proof of the last fact, due to J. Kürschák [Honsberger, 1985].

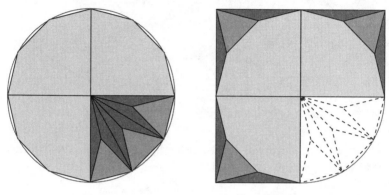

FIGURE **13.8**

Readers interested in plane dissections and their mathematical properties will find the books by G. Frederickson [Frederickson, 1997 and 2002] of interest.

13.4   Challenges

13.1   Figure 13.9 is a drawing of the "sedan chair" in Sam Loyd's famous puzzle (Figure 13.1). Divide the sedan chair into the minimum number of pieces to form a perfect square. [Hint: the minimum is two!]

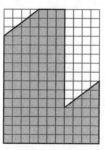

FIGURE **13.9**

13.2 Verify the claim following Figure 13.6 that Loyd's "solution" to the Smart Alec puzzle is faulty.

13.3 Show how the "L" shaped region in Figure 13.10 formed from three-quarters of a square can be dissected into n congruent pieces, each similar to the original "L," for $n = 9, 16$, and 25.

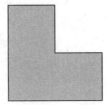

FIGURE **13.10**

13.4 Show how the trapezoid in Figure 13.11 formed from three equilateral triangles can be dissected into n congruent pieces, each similar to the original trapezoid, for $n = 4, 9, 16$, and 25.

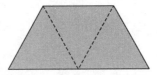

FIGURE **13.11**

13.5 Given two equilateral triangles with sides of 3 cm and 4 cm, dissect the triangles to form one equilateral triangle with 5 cm sides.

14

Moving Frames

Drawing an approximate graph of a function has been, for decades, a common exercise on visual representation in mathematics classes. While hand-made graphs on paper or a blackboard have always been a rather tedious procedure, today's calculators and computers help us to draw sophisticated graphs (and to work with them) rapidly and efficiently.

FIGURE 14.1

We assume the reader is already familiar with the use of such devices for graphing functions. So in this chapter we focus our attention on showing how some important functional properties may be better understood with visualization.

14.1 Functional composition

The sum $f + g$ and the product $f \cdot g$ of functions f and g are pointwise operations where to each number a one assigns $f(a) + g(a)$ or $f(a) \cdot g(a)$, respectively. The most powerful operation among functions is composition. Let f and g be two functions such that the range of g is a subset of the domain of f, so that it is possible to consider the composite function $f \circ g$, defined by $(f \circ g)(a) = f(g(a))$. Our visual challenge here is: given graphs of f and g, can we construct the graph of $f \circ g$? Note that this problem includes as a special case graphing the iterates of a function.

The procedure we outline in Figure 14.2 appears in [Menger, 1952]. Start with a point a, use the graph of $y = g(x)$ to find $g(a)$ on the y-axis, then by means of the graph $y = x$ we can relocate $g(a)$ on the x-axis. Now use the graph of $y = f(x)$ to locate $f(g(a))$. We now translate this value $f(g(a))$ horizontally to lie above $(a, 0)$, and mark the desired final point $(a, f(g(a)))$. As the picture illustrates, there is a "moving frame" with one corner on the diagonal $y = x$, the two adjacent corners on the graphs of f and g, and the opposite corner on the graph of $f \circ g$.

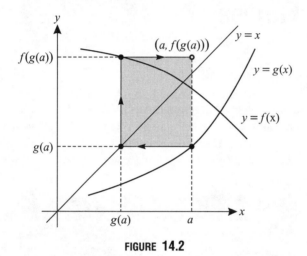

FIGURE 14.2

In particular, we can use this idea to find g^{-1} for a given one-to-one function g. Since $(g^{-1} \circ g)(x) = x$ for all x in the domain of g, we have $(a, g^{-1}(g(a))) = (a, a)$, and hence g^{-1} must be the function such that the moving frame has both its southwest and northeast corners on the diagonal. So the frame must be a square and therefore g^{-1} must be the function symmetric to g with respect to the diagonal $y = x$.

14.2   The Lipschitz condition

A function f is said to satisfy a *Lipschitz condition* if there exists a constant $M > 0$ such that

$$|f(x_2) - f(x_1)| \le M \, |x_2 - x_1|$$

for all x_1, x_2 in the domain of f. This is a global property affecting the entire graph of f and implies the continuity of the function. To "see" a Lipschitz condition, we consider a pair of lines $y = Mx$ and $y = -Mx$, an idea due to Miguel de Guzmán [de Guzmán, 1996]. The above condition, in the form

$$\left| \frac{f(x_2) - f(x_1)}{x_2 - x_1} \right| \le M$$

may be interpreted as follows: at each point $(a, f(a))$ the entire graph of f lies between the lines $y = f(a) + M(x - a)$ and $y = f(a) - M(x - a)$. So as the graphs of $y = Mx$ and $y = -Mx$ move along the graph of f (with the point of intersection of the two lines on the graph) the graph of f always lies in the unshaded region between these lines, as illustrated below.

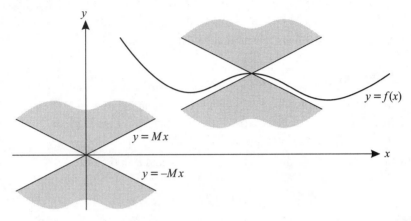

FIGURE **14.3**

Pedagogical remark. A transparency with the lines $y = \pm Mx$ moving along the graph of a function (either projected on a screen or on a chalkboard) illustrates a Lipschitz condition.

14.3 Uniform continuity

Let f be a continuous function and assume that for any $\varepsilon > 0$ we can find a $\delta > 0$ and construct a rectangle $[-\delta, \delta] \times [-\varepsilon, \varepsilon]$ such that whenever we move this rectangular window so that it is centered at $(a, f(a))$, all points on the graph of f for x within δ units of a lie inside the window.

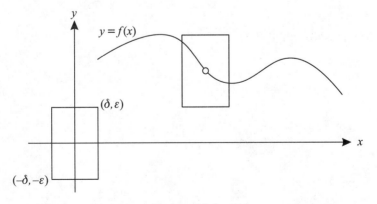

FIGURE **14.4**

When this very special situation (the width of the window depends only on its height, and not on the point a where it is centered) arises, we say that f is *uniformly continuous* on its domain.

Remark: Continuity versus uniform continuity   The notion of *continuity* is strictly local: one says that f is continuous at a point a if for any $\varepsilon > 0$ there exists a $\delta > 0$ such that whenever $|x - a| < \delta$ we have $|f(x) - f(a)| < \varepsilon$. In terms of the moving window described in the above paragraph, the width of the window in this case may well depend on both its height and the point a where the window is centered.

14.4   Challenges

14.1 a) Modify the functional composition procedure in Section 14.1 to show how to illustrate functional iteration, i.e., given a function $y = f(x)$ and a number a, to locate $f(a)$, $f(f(a))$, $f(f(f(a)))$, etc.

   b) In particular, show that if $f(x) = \sqrt{2 + x}$ and $a = 0$, then

$$\sqrt{2 + \sqrt{2 + \sqrt{2 + \cdots}}} = 2.$$

14.2 A function $f : [0, \infty) \to [0, \infty)$ is said to be *subadditive* if $f(x + y) \leq f(x) + f(y)$, for all x, y in the domain of f. Find a visual interpretation of this condition.

14.3 A continuous function f is said to be *convex* if

$$f\left(\frac{x + y}{2}\right) \leq \frac{f(x) + f(y)}{2}$$

for all x, y in the domain of f. What does convexity mean from a visual point of view?

14.4 Let f be a continuous function. What kind of moving frame would you design to check for the possible monotonicity of the function?

15

![gray bar]

Iterative Procedures

Closely related to the idea of using multiple copies of a picture (see Chapter 11) is the following procedure wherein we employ multiple (indeed, sometimes infinitely many) copies of a picture, but in such a way so that part of the picture is a scaled version of the entire picture. For example, in Figure 15.1(a) the northeast quarter of the square is a scaled version of the entire square. If we now let the side length of the largest square be 1 and label the interior rectangles and squares with their respective areas, we have in Figure 15.1(b) a visual proof of the sum of the geometric series with common ratio $1/2$: $1/2 + 1/4 + 1/8 + \cdots = 1$.

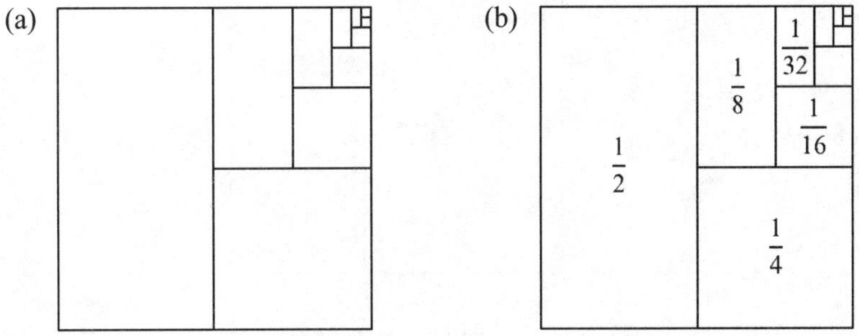

FIGURE **15.1**

We explore this idea in this chapter.

15.1 Geometric series

Here are two pictures [Mabry, 1999; Ajose, 1994] that illustrate that the sum of the series $1/4 + (1/4)^2 + (1/4)^3 + \cdots$ is $1/3$. In Figure 15.2(a), note that the largest dark gray triangle is $1/4$ of the original triangle, the next largest is $1/4$ of $1/4$ of the original triangle, and so on; and that together the dark gray triangles make up $1/3$ of the original

(a)   (b)

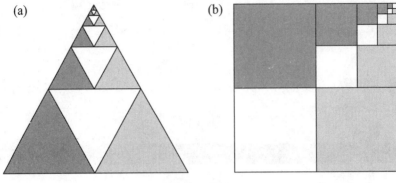

FIGURE **15.2**

triangle (as do the white and the light gray triangles). In Figure 15.2(b) we accomplish the same objective with squares.

This idea can be extended to find the formula for the sum of a general geometric series (with a positive first term a and common ratio r) [Bivens and Klein, 1988]. In Figure 15.3, the iterative procedure consists in partitioning the large white triangle into similar trapezoids. The gray triangle and the large white triangle are similar; hence the ratio of horizontal to vertical sides in each is the same, which yields

$$\frac{a + ar + ar^2 + ar^3 + \cdots}{1} = \frac{a}{1 - r}.$$

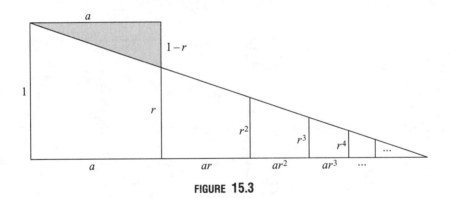

FIGURE **15.3**

15.2 Growing a figure iteratively

Rather than subdivide a given figure into successively smaller and smaller copies of itself, we can *enlarge* a figure in an iterative fashion. In Figure 15.4(a) we have a square array of $4^2 = 16$ dots, which "grows" to an array of $4^3 = 64$ dots in Figure 15.4(b) by the addition of three copies of itself along the top and right side of the figure. In turn, Figure 15.4(b) "grows" to an array of $4^4 = 256$ dots in Figure 15.4(c) in the same manner [Sher, 1997].

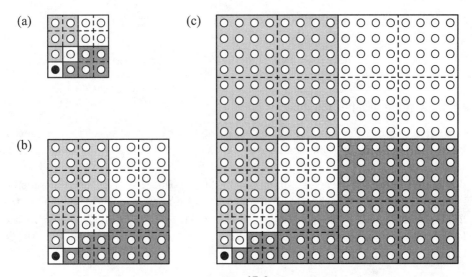

FIGURE 15.4

After n steps, the side of the square has length $1 + 1 + 2 + 4 + \cdots + 2^n = 2^{n+1}$, and computing the number of dots in the square in two ways yields

$$1 + 3 \left(1 + 4 + 4^2 + \cdots + 4^n \right) = \left(2^{n+1} \right)^2 = 4^{n+1},$$

and hence

$$1 + 4 + 4^2 + \cdots + 4^n = \frac{4^{n+1} - 1}{3}.$$

A similar iterative growth pattern can be employed to examine sums of triangular numbers $T_n = 1 + 2 + \cdots + n$ (introduced in Section 1.2) when n is a power of 2. For example, the right-most part of Figure 15.5 shows that

$$3 \left(T_1 + T_2 + T_4 + T_8 \right) + 3 = T_{17}, \quad \text{or} \quad T_1 + T_2 + T_4 + T_8 = \frac{1}{3} T_{17} - 1.$$

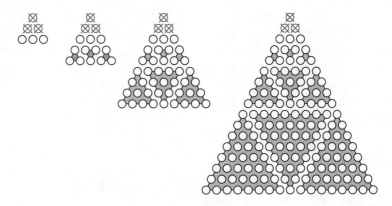

FIGURE 15.5

Continuing in this manner, we have

$$\sum_{k=0}^{n} T_{2^k} = T_1 + T_2 + T_4 + \cdots + T_{2^n} = \frac{1}{3}T_{2^{n+1}+1} - 1.$$

15.3 A curve without tangents

In nearly every calculus text one encounters examples of curves that fail to have tangent lines at particular points. Common examples are the graph of the absolute value function $y = |x|$ and the curve given by $y = 0$ at $x = 0$ and $y = x\sin(1/x)$ for $x \neq 0$. At the origin each of these curves fails to have a tangent line. However, using iteration it is easy to construct a curve that fails to have a tangent at every point of the curve! The following example is from Karl Menger's 1952 booklet *You Will Like Geometry*, written to accompany an interactive geometry exhibit at the Museum of Science and Industry in Chicago, and reprinted in [Schweizer et al., 2003].

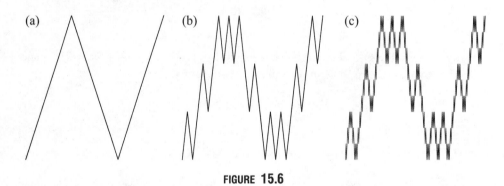

FIGURE 15.6

In part (a) of Figure 15.6 we have a curve in the shape of the letter N. In part (b), each of the three lines of the N in (a) has been replaced by 3 Ns, together forming a zigzag line of 9 Ns. In part (c), each of the 9 Ns in (b) has been replaced by 9 Ns, for a total of 81 Ns. If we iterate this pattern, at each stage we obtain a more jittery line, and in the limit we obtain a curve without a tangent at any point. As Menger noted in his booklet, the first curve without any tangent was discovered by Weierstrass in about 1870. At that time it was considered exceptional, but today it is known that there are more curves without any tangents than there are smooth curves with tangents.

Similar iterative procedures are commonly employed to construct fractals with such intriguing names as the Sierpinski carpet, the Barnsley fern, the Koch snowflake, and the Menger sponge.

15.4 Challenges

15.1 Use pictures similar to Figures 15.1 and 15.2 to illustrate
(a) $1/3 + (1/3)^2 + (1/3)^3 + \cdots = 1/2$,
(b) $1/5 + (1/5)^2 + (1/5)^3 + \cdots = 1/4$.

15.2 What series (and their sums) are illustrated by these pictures?

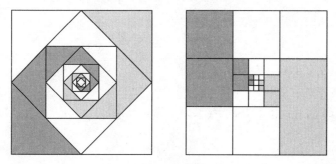

FIGURE 15.7

15.3 Use the sequence of pictures below [Sher, 1997] to illustrate that

$$1 + 3 + 3^2 + \cdots + 3^n = \frac{3^{n+1} - 1}{2}.$$

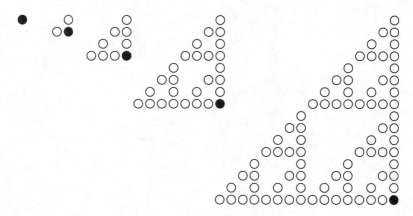

FIGURE 15.8

15.4 Using the fact that the golden ratio $\phi = \left(1 + \sqrt{5}\right)/2$ satisfies $\phi^2 = \phi + 1$, devise an iterative procedure with compass and straightedge to construct segments of lengths ϕ^n for $n \geq 1$.

15.5 With an illustration similar to Figure 15.5, show that

$$T_1 + T_3 + T_7 + \cdots + T_{2^n-1} = \frac{1}{3} T_{2^{n+1}-2}.$$

16

Introducing Colors

We often introduce color in mathematical pictures for aesthetic reasons or to distinguish various parts of the picture. In this chapter we illustrate the use of color (or just black and white, or various shades of gray or patterns) to enhance an argument. This idea is especially useful when working with tilings (tilings were introduced in Chapter 10).

16.1 Domino tilings

Given a standard 8×8 checkerboard (see Figure 16.1(a)), it is a simple matter to "tile" the board with 32 1×2 dominoes of the appropriate size (that is, place the dominoes on the board so that they don't overlap and every square is covered), as each domino covers exactly two squares. Indeed, any $2n \times 2n$ checkerboard can be similarly tiled.

But what if we remove two opposite corner squares from the 8×8 checkerboard, as illustrated in Figure 16.1(b)—can the resulting "deficient" checkerboard be tiled with 31 dominoes? The answer is no, since 31 dominoes will cover exactly 31 light gray and 31 dark gray squares, and our deficient checkerboard has 32 light gray and 30 dark gray squares (both of the removed squares were dark gray).

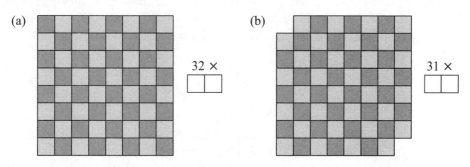

FIGURE **16.1**

16.2 *L*-Tetromino tilings

An *L-tetromino* is a tile in the shape of the letter "L," formed from four squares as shown in Figure 16.2(a). Clearly the 8×8 checkerboard can be tiled by 16 *L*-tetrominoes, since two *L*-tetrominoes cover a 2×4 board, as shown in Figure 16.2(b). But if we remove a 2×2 square from anywhere on the board (such as illustrated in Figure 16.2(c)), can this deficient checkerboard be tiled with 15 *L*-tetrominoes?

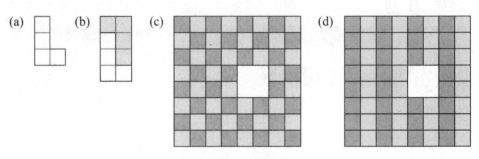

FIGURE **16.2**

With the "standard" coloring of the checkerboard, each *L*-tetromino, no matter its orientation, covers two light gray and two dark gray squares, and the deficient checkerboard has 30 squares of each color, hence we do not have a contradiction (nor do we have a proof that the deficient checkerboard *can* be tiled!). However, suppose we color the board as shown in Figure 16.2(d). Again, we have 30 squares of each color, but each *L*-tetromino, no matter its orientation, will cover one light gray and three dark gray squares (class A) or one dark gray and three light gray squares (class B). If the tiling is possible, then we would have x class A *L*-tetrominoes and y class B *L*-tetrominoes, where $4x + 4y = 60$ (since each of the 60 squares would be covered) and $3x + y = 30$ (since each of the 30 dark gray squares would be covered). But the unique solution to this pair of equations is $x = y = 7.5$, and hence the deficient checkerboard in Figure 16.2 cannot be covered by 15 *L*-tetrominoes.

16.3 Alternating sums of triangular numbers

In Section 1.3, we used colors to show that alternating sums of squares are signed triangular numbers. In a similar fashion, we can use colors to show that an alternating sum of triangular numbers is a square, that is, if $T_k = 1 + 2 + \cdots + k$, then

$$T_1 - T_2 + T_3 - \cdots + T_{2n-1} = n^2.$$

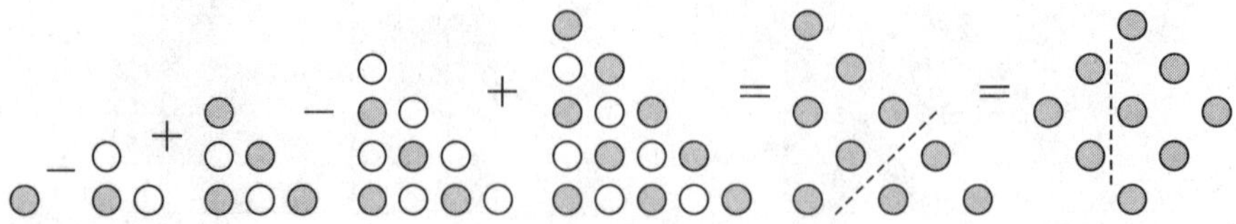

FIGURE **16.3**

16.4   In space, four colors are not enough

One of the best-known theorems related to colors is the celebrated Four Color Theorem, which states that four or fewer colors suffice to color any (suitably defined) map in the plane so that adjacent countries have different colors. Appel and Haken proved this theorem in 1976, following a century of attempts by numerous professional and amateur mathematicians. What happens in higher dimensions? If four is the "chromatic number" of the plane, what is the chromatic number of space? The following sequence of pictures shows that no finite number of colors is sufficient to color a three-dimensional "map":

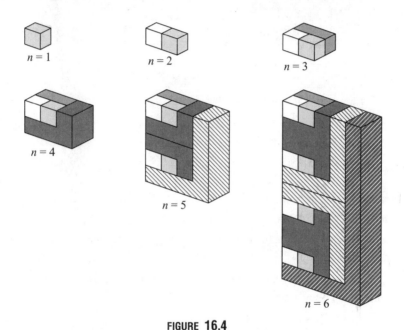

FIGURE 16.4

16.5   Challenges

16.1 A *T-tetromino* is formed from four squares in the shape of the letter "T," as illustrated at the right.

FIGURE 16.5

    a) Can the standard 8×8 checkerboard in Figure 16.1(a) be tiled with 16 *T*-tetrominoes?

    b) Can the deficient 8×8 checkerboard in Figure 16.2(c) be tiled with 15 *T*-tetrominoes?

16.2 There are two types of *trominoes*, the *straight tromino* and the *L-tromino*, illustrated at the right. Since the standard 8×8 checkerboard has 64 squares, it clearly can't be covered by any set of trominoes. However, if we remove one square, the resulting deficient checkerboard has 63 squares.

FIGURE 16.6

a) Can it be covered by 21 straight trominoes? [Hint: The answer depends upon *which square* is removed! Consider the two colorings below.]

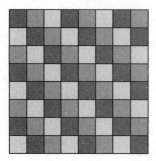 

FIGURE **16.7**

b) What about $n \times n$ checkerboards for other values of n (not a multiple of 3)?

c) Show that no matter which square is removed from the 8×8 checkerboard, the resulting deficient checkerboard can be tiled with 21 L-trominoes. Indeed, any $2^n \times 2^n$ checkerboard with one square removed can be tiled with L-trominoes [Golomb, 1954].

16.3 Show that two colors suffice to color any partition of the plane generated by a finite number of lines.

16.4 How many colors are needed to color a partition of the plane generated by circles rather than lines?

16.5 For which values of m and n can a rectangular $m \times n$ checkerboard be tiled with copies of the L-tetromino in Figure 16.2(a)?

17

Visualization by Inclusion

This technique is especially powerful for proving numerical inequalities between positive numbers. Its secret consists of the basic fact that whenever a set A is included in another set B, necessarily any measure of A (cardinality, length, area, volume, ...) will be less than (or equal to) the corresponding measure of B.

17.1   The genuine triangle inequality

Given any triple a, b, c of positive numbers, there exists a triangle with sides of lengths a, b, and c if and only if $a + b > c$, $b + c > a$, and $c + a > b$. Without loss of generality, assume $a \leq b \leq c$. Then only the first inequality above ($a + b > c$) is not trivial.

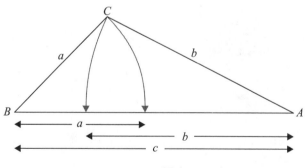

FIGURE **17.1**

In a triangle with side lengths a, b, c, rotating sides BC and AC to coincide with AB, we see that the side AB is included within the union of sides BC and AC, hence $c < a + b$. The converse is immediate.

An interesting consequence is that the square root function is subadditive [recall Challenge 14.2: a function f is subadditive if $f(a + b) \leq f(a) + f(b)$]. Given positive numbers a and b, construct a right triangle with legs $\sqrt{a}$ and $\sqrt{b}$, and hypotenuse $\sqrt{a+b}$ (you may confirm this with the Pythagorean theorem). Then the above triangle inequality yields $\sqrt{a} + \sqrt{b} > \sqrt{a+b}$.

17.2 The mean of the squares exceeds the square of the mean

Given a pair of positive numbers, how does the average of the squares compare to the square of the average? The answer is revealed in Figure 17.2 and a simple computation.

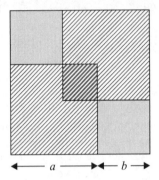

$$2a^2 + 2b^2 \geq (a+b)^2$$

$$\therefore \frac{a^2 + b^2}{2} \geq \left(\frac{a+b}{2}\right)^2$$

FIGURE 17.2

Indeed, the result can be extended to any finite number n of positive numbers, as shown in Figure 17.3 below for $n = 4$ [Nelsen, 2000b].

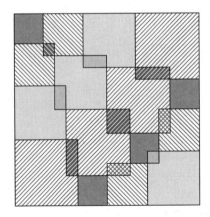

$$n\left(a_1^2 + a_2^2 + \cdots + a_n^2\right) \geq (a_1 + a_2 + \cdots + a_n)^2$$

$$\therefore \frac{a_1^2 + a_2^2 + \cdots + a_n^2}{n} \geq \left(\frac{a_1 + a_2 + \cdots + a_n}{n}\right)^2$$

FIGURE 17.3

17.3 The arithmetic mean-geometric mean inequality for three numbers

Let's establish the arithmetic mean-geometric mean inequality for three numbers, i.e., $\sqrt[3]{xyz} \leq (x + y + z)/3$ for $x, y, z > 0$ [the inequality for two numbers was encountered in Section 3.4]. To do so, we first introduce a change of variables, $x = a^3$, $y = b^3$,

$z = c^3$, and show that $3abc \leq a^3 + b^3 + c^3$. We'll do it in two steps [Alsina, 2000b], first proving a lemma.

Lemma. $ab + bc + ac \leq a^2 + b^2 + c^2$.

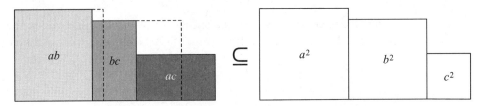

FIGURE 17.4

The image in Figure 17.4 illustrates a typical case of inclusion, since for $a > b > c$, the rectangles with areas ab, bc, and ac are contained in the union of the squares of areas a^2, b^2, and c^2.

Theorem. $3abc \leq a^3 + b^3 + c^3$.

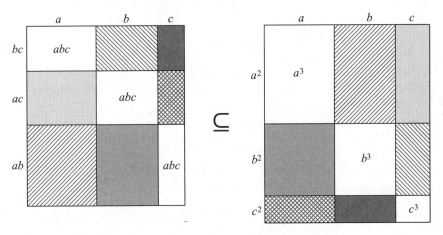

FIGURE 17.5

By the lemma, the rectangle on the left is included in the one on the right (note that both have the same base $a + b + c$). In the rectangle on the left we see three rectangles each with area abc plus six shaded rectangular regions. On the right we see six shaded rectangles with the same areas as the ones on the left along with three rectangles on the diagonal with areas a^3, b^3, and c^3. In this picture (as opposed to Figure 17.4) a^2, b^2, c^2, ab, bc, and ac represent lengths rather than areas.

Figure 17.6 illustrates the same inequality $3abc \leq a^3 + b^3 + c^3$ but in the form $abc \leq \frac{1}{3}a^2 \cdot a + \frac{1}{3}b^2 \cdot b + \frac{1}{3}c^2 \cdot c$.

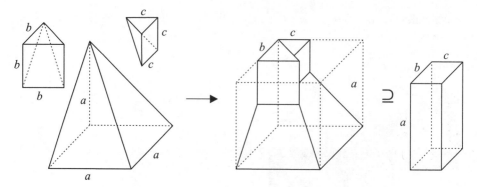

FIGURE **17.6**

Here abc represents the volume of the box with sides of length $a \geq b \geq c$. But this box is clearly contained in the union of the three right pyramids whose bases are squares of areas a^2, b^2, and c^2, and with heights a, b, and c, respectively.

17.4   Challenges

17.1 In the first paragraph of Section 3.4 we described a proof of the fact that the sum of a positive number and its reciprocal is always at least 2. Use Figure 17.7 to give another proof.

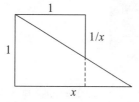

FIGURE **17.7**

17.2 Use inclusion to illustrate *Jordan's inequality*: $\dfrac{2x}{\pi} \leq \sin x \leq x$ for $0 \leq x \leq \dfrac{\pi}{2}$.

17.2 Use Figure 17.2 to establish the following inequalities for positive a and b:

   a) the root mean square-arithmetic mean inequality: $\sqrt{\dfrac{a^2 + b^2}{2}} \geq \dfrac{a + b}{2}$;

   b) the arithmetic mean-geometric mean inequality: $\dfrac{a + b}{2} \geq \sqrt{ab}$; and

   c) the geometric mean-harmonic mean inequality: $\sqrt{ab} \geq \dfrac{2ab}{a + b}$.

   [Hints for b) and c)): replace (a, b) by $(\sqrt{a}, \sqrt{b})$ and $(1/\sqrt{a}, 1/\sqrt{b})$, respectively.]

17.4 Show that the results in Section 17.2 hold for any real numbers.

17.5 Use inclusion to establish the *mediant property* for positive numbers a, b, c, d: if $a/b < c/d$, then $a/b < (a + c)/(b + d) < c/d$ (recall Section 2.2).

18

Ingenuity in 3D

The aim of this chapter is to use concrete examples to show how some geometrical problems can be readily solved in three dimensions by means of ingenious "hands-on" strategies, while it would be almost impossible (or extremely tedious) to address the same problems in the traditional way.

18.1   From 3D with love

We start with a collection of problems that motivate the development of three-dimensional strategies.

Problem 1 *Suppose we are sitting at a table, on which there is a box of unknown dimensions and a tape measure. What is the easiest way to determine the length of the diagonal of the box?*

The key to the solution is the table! We place the box on the corner and using the tape measure we move the box along one side of the table until we can measure directly the diagonal in the "empty box space" determined by the corner of the table and the corresponding real vertex of the box.

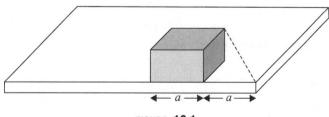

FIGURE 18.1

Problem 2 *Can we create a line segment of length (exactly) 2π?*

The answer is yes if we have the appropriate three-dimensional objects! Take a cylinder and choose as the unit of measure its radius. Rolling the cylinder one complete revolution on a plane yields a line segment of length 2π. This, of course, is impossible to do with only a straightedge and compass.

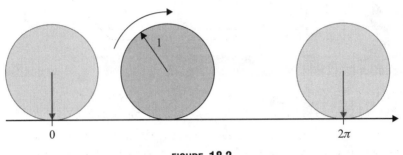

0 2π

FIGURE **18.2**

Problem 3   *Take six pencils all the same length. Can we arrange them in such a way that each one touches all of the others? What about seven pencils?*

The answer to both problems is yes and here are solutions. Note that in the case of seven pencils, one pencil is placed vertically at the center.

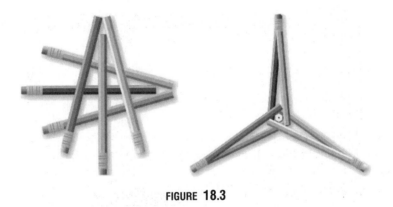

FIGURE **18.3**

If you allow pencils of different lengths, can you arrange *eight* pencils so that each is touching all the others? See Challenge 18.4.

Problem 4   *Take a rectangular piece of cardboard, and hold it with one vertex on the surface of a table, as shown in Figure 18.4. What is the relationship among the heights of the other three vertices?*

The vertex opposite the one touching the table has a height equal to the sum of the heights of the other two vertices. Think of moving along the perimeter of the rectangle from the vertex at the bottom to the vertex at the top.

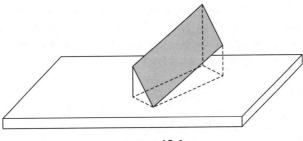

FIGURE **18.4**

Problem 5 *Suppose we have a ball (e.g., a basketball, soccer ball, billiard ball) and a ruler in a classroom. How can we accurately measure the radius of the ball?*

In a classroom, of course, we can find some chalk, so mark a spot on the surface of the ball with the chalk and place the ball in one corner of the room so that the ball touches the floor and the chalk spot touches a wall, marking the wall with the chalk at a height equal to its radius. Now using the ruler we simply measure the height of the chalk mark on the wall.

Problem 6 *Suppose we have a ruler, paper and scissors. How can we construct a cylinder with a helix of a given slope on its surface?* [The *slope* of a helix is the ratio of the vertical distance between loops to the circumference of the cylinder.]

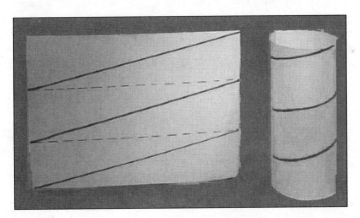

FIGURE **18.5**

The key is in the definition of the slope, and Figure 18.5 answers the Problem. Also note that the shortest path between two points on a cylinder (when one point is not vertically above the other) is a fractional turn of a helix.

18.2 Folding and cutting paper

Constructing objects by cutting and folding paper has been a common activity since paper was invented. In the literature there is an extensive collection of guides for paper-folding addicts. Perhaps the most elegant constructions are those based on the Japanese tradition known as Origami.

From a pedagogical point of view, many paper constructions have mathematical interest. In this section we'll present some surprising exercises based on cutting and folding paper.

Problem 7   *Take a sheet of paper and scissors. Can you cut a hole in the paper large enough to walk through*

Clearly there is no way to walk through a "standard hole" in the paper. But if we interpret as a "hole" the space bounded by a continuous piece of the sheet of paper, then our challenge has infinitely many solutions. Here we present one:

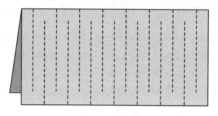

FIGURE **18.6**

Fold the paper in half and cut on the dashed lines (through both halves of the folded paper) as indicated in Figure 18.6. When opened up you will have a rather large hole to walk through.

Problem 8   *Figure 18.7(a) illustrates an "S" shaped piece of paper with a "flap" protruding from the front. Can it be constructed from a single sheet of paper using only scissors?*

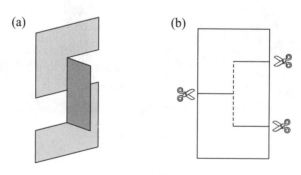

FIGURE **18.7**

Yes! Cut the sheet of paper as indicated in Figure 18.7(b), then make a valley fold on the upper half of the dashed line and a mountain fold on the lower half [Tanton, 2001a].

Problem 9   *Take a rectangular sheet of paper and by folding from one corner mark the largest inscribed square. Do the same with the remaining rectangular piece. Thus you*

have marked another smaller square and a final rectangular piece. Can this be done
with any rectangle? What must be the shape of the original rectangle so that this final
rectangle will be similar to the original rectangle?

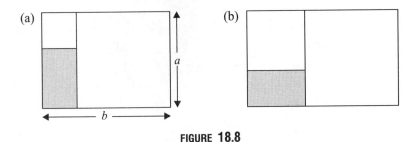

FIGURE **18.8**

If the original rectangle measures $a \times b$, with $b > a$, then the squares have side lengths
a and $b - a$, so that the dimensions of the shaded rectangle are $2a - b \times b - a$. Thus
the construction is only possible when $b < 2a$. If the shaded rectangle has the opposite
orientation as the original (as illustrated in Figure 18.8(a)), then $(2a - b)/(b - a) = b/a$,
and hence $b/a = \sqrt{2}$. If the shaded rectangle has the same orientation as the original (as
illustrated in Figure 18.8(b)), then $(b - a)/(2a - b) = b/a$, and hence $b/a = \phi \simeq 1.618$,
the golden ratio.

Problem 10 *Start with two congruent squares of paper, A and B. The one on the top,*
A, covers 1/4 of the area of B. Now rotate A about the center of B through any angle.
How does the fraction of B covered by A change?

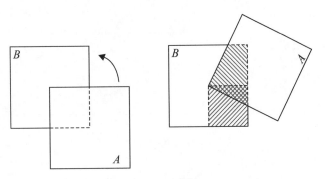

FIGURE **18.9**

The fraction of B covered by A remains constant at $1/4$, as can be seen by comparing
the areas of the shaded regions. What would happen if we used congruent rectangles
rather than squares? What about equilateral triangles?

Problem 11 *Suppose we have two identical sheets of paper. With one, we glue the*
parallel long sides together to form a cylinder. We want to do the same with the other
sheet but before we do we want to make a hole in it in such a way that we can insert

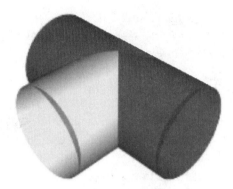

FIGURE **18.10**

the first one into the hole to have a three-dimensional "T". Find the shape of the hole that we need to make. The two cylinders are joined as shown in Figure 18.10.

If one intersects a circular cylinder with a plane (which also intersects the axis of the cylinder), one obtains an elliptical section, as shown in Figure 18.11(a). But what does the intersection look like if we unroll the cylinder? It certainly *looks* like a sine wave in Figure 18.11(b), and so it is.

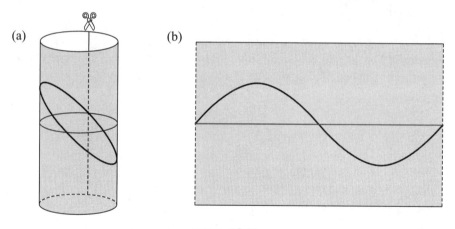

FIGURE **18.11**

In xyz-coordinates, an equation of the cylinder is $x^2 + y^2 = 1$, and of the plane $z = Ay$. Hence a parametric representation of the curve of intersection is $(\cos\theta, \sin\theta, A\sin\theta)$ for θ in $[0, 2\pi]$. Unrolling the cylinder is equivalent to looking at the curve in a θ-z plane, thus an equation of the curve is $z = A\sin\theta$.

On the left in Figure 18.12 we see the classical architectural form known as a *groin vault,* formed by the intersection of two identical half cylinders whose axes are perpendicular. On the right is a drawing of the region in the vault common to the two cylinders. In order to construct a paper model of the region, we use the previous observation about elliptical sections and sine curves.

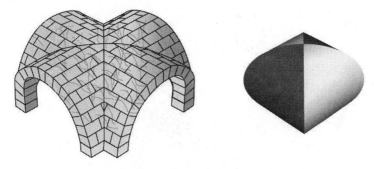

FIGURE **18.12**

In Figure 18.13(a) we have a template for folding a model of the region under the groin vault. The curves are portions (1/4) of a sine wave. Returning to Challenge 11, we see that the solution consists of half of a sine wave and its reflection, as shown in Figure 18.13(b).

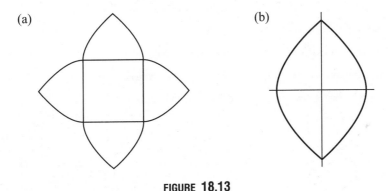

(a)   (b)

FIGURE **18.13**

Problem 12 *Making friezes.*

A frieze is a design or pattern that repeats itself in one direction (which we will take to be horizontal). Friezes are often seen as decoration in architecture. A mathematical analysis reveals that there are seven different frieze patterns possible, as shown in Figure 18.14.

Recall that isometries in the plane (see Chapter 6) are transformations that preserve distances. Formally, they are of the form $T_\mathbf{x} \circ f$, where $T_\mathbf{x}$ denotes a translation generated by a vector $\mathbf{x}$ and f is either a rotation around the origin or a symmetry with respect to a horizontal or vertical line through the origin. Thus one can associate with any figure F in the plane its symmetry group $S(F)$ defined as the set of plane isometries which leave F invariant. The figure F is called a frieze whenever there is a line in the plane which is invariant by all transformations in $S(F)$ and there is a vector $\mathbf{x} \neq \mathbf{0}$ such that all translations in $S(F)$ have the form $T_{n\mathbf{x}}$ for some integer n.

FIGURE **18.14**

Our aim in this section is to discuss how to take strips of paper and scissors, and make models of the seven friezes as an alternative to the above drawings. To this end we need to combine three basic moves:

(i)   **The horizontal fold** to obtain horizontal symmetry;

(ii) **The accordion fold** to introduce vertical reflections; and

(iii) **The cylinder effect** to translate figures. See Figure 18.15.

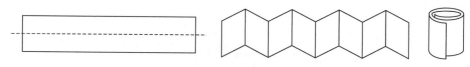

FIGURE **18.15**

With these three operations we can make paper friezes by using scissors to cut a pattern into a rolled or folded strip of paper and then unroll or unfold the strip. For example, if we roll a strip into a cylinder and cut a pattern out of the paper at the top of the cylinder and unroll it, we will get a pattern with the same symmetry as the first frieze in Figure 18.14. If we also cut out a pattern at the bottom of the cylinder symmetric to the one at the top with respect to the center of the cylinder, we will obtain a pattern with the symmetry of the second frieze in Figure 18.14. Other patterns can be obtained from strips with the accordion and/or horizontal folds.

18.3   Unfolding polyhedra

Sometimes three-dimensional problems admit planar solutions. A classic example appeared in Section 9.4, where we searched for a path of shortest length between a spider

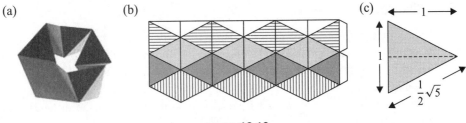

(a) (b) (c)

FIGURE **18.16**

and a fly in a rectangular room. The solution resulted from a "3D→2D→3D" strategy: unfold the room to form a two-dimensional figure, find the shortest path there, then return to the three-dimensional figure.

Part of the strategy in the minimal path on a cube problem concerned finding the appropriate way to unfold the cube (also see Challenge 19.1 in the next chapter). Folding and unfolding tetrahedra (triangular pyramids) leads to an interesting three-dimensional structure called a *kaleidocycle*. A kaleidocycle is a chain of six, eight, or even ten tetrahedra joined at their edges that flex and rotate to show different sides, as illustrated in Figure 18.16(a).

The template in Figure 18.16(b) can be folded to form six joined tetrahedra, each with different patterns on its four faces (each column of triangles forms a tetrahedron, the white triangles and tabs are for gluing the structure together). Hint: make valley folds on the vertical lines and mountain folds on the slanted lines!

The tetrahedra in the kaleidocycle in Figure 18.16 are not regular, i.e., the faces are not equilateral triangles. The faces are isosceles triangles with the shape indicated in Figure 18.16(c). The reason for this shape will be apparent when you construct your first kaleidocycle. Adding triangles to the template will result in kaleidocycles with eight or ten tetrahedra, in these cases equilateral triangles may be used.

It is interesting to note that as these rings of tetrahedra move, their volumes and surface areas remain constant, while distances between points change—that is, invariance of areas and volumes does not imply metric invariance. All the figures in the picture below have this property of motion, breaking the traditional rigidity of "classical" shapes.

FIGURE **18.17**

18.4 Challenges

18.1 Take a long strip of paper, give the paper a half twist and glue or tape the ends together—the result is called a *Möbius strip* or *Möbius band*. If you were to cut the strip down the middle length-wise, what do you think you would get? Do it, you may be surprised by the result. What do you think would happen with two or more half-twists in the strip? What about cutting the strip $1/3$ the distance from one edge?

18.2 One magazine A lies on the top of another one B, as indicated in Figure 18.18. Does A cover more or less than half of the area of B?

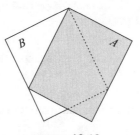

FIGURE **18.18**

18.3 Suppose you have a rectangular sheet of paper that measures $a \times b$ with $a < b$. You may join the longer sides (of length b) or the shorter sides (length a) to obtain, in each case, a cylinder. Clearly the cylinders have equal lateral surface areas. Are the volumes the same?

18.4 Show how to arrange eight pencils so that each one touches all of the others.

19

Using 3D Models

Many mathematical properties of three-dimensional objects may be seen easily by making appropriate three-dimensional models. A large collection of physical resources for the construction of models is presented in Part II.

19.1 Platonic secrets

There are precisely five Platonic solids—polyhedra whose faces are congruent regular polygons and where the same number of faces meet at each vertex. They are illustrated in Figure 19.1, and their names and descriptions are: the *tetrahedron* (four triangular faces); the *cube* (six square faces); the *octahedron* (eight triangular faces); the *dodecahedron* (twelve pentagonal faces); and the *icosahedron* (twenty triangular faces). To be consistent, the cube should be called a *hexahedron*.

FIGURE **19.1**

Each of the five Platonic solids has many intriguing properties, some of which are not well known. In this section we describe for each an intriguing "secret."

The dodecahedron's secret: a cube with six roofs

The dodecahedron and the cube—the two solids on the left in Figure 19.1—have the number twelve in common. The cube has twelve edges, the dodecahedron has twelve faces, and in fact a cube can be inscribed in a dodecahedron in such a way that each edge of the cube is a diagonal of a pentagonal face of the dodecahedron. See Figure 19.2(a). Consequently, the dodecahedron can be thought of as the union of a cube with six roofs, one on each face, as illustrated in Figure 19.2(b).

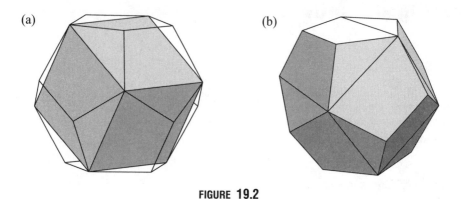

(a)   (b)

FIGURE 19.2

Let's construct a model of this relationship between the cube and the dodecahedron. First construct a cube of side length s and then cut out six copies of Figure 19.3, consisting of a square of side s adjoined to a pentagon whose side also has length s (hence the side length of the pentagonal face of the dodecahedron will be $(\phi - 1)s \simeq 0.618s$, see Section 7.2). Cutting, folding and gluing each piece as indicated we will form a roof with a square base.

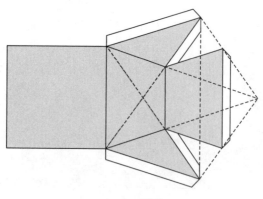

FIGURE 19.3

After making the six roofs (it's teamwork!) place them on the cube as in Figure 19.2(b) so that the dodecahedron will appear with its twelve pentagonal faces, each one with one diagonal corresponding to a side of the interior cube.

The icosahedron's secret: three golden rectangles inside

The icosahedron is one of the three Platonic solids made up of triangles, the other two being the tetrahedron and the octahedron. Convex polyhedra all of whose faces are congruent equilateral triangles are called *deltahedra* (from the Greek letter delta, Δ). Figure 19.4 illustrates the eight different deltahedra.

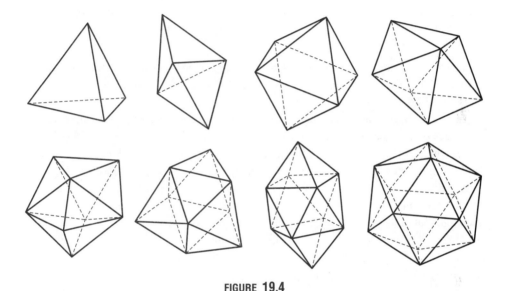

FIGURE **19.4**

The icosahedron is the union of two pentagonal pyramids and a pentagonal antiprism, as illustrated in Figure 19.5 (an *antiprism* is a polyhedron constructed from two regular *n*-gons and 2*n* equilateral triangles). Since the icosahedron has a pentagonal section (in black in Figure 19.5(c)), it is not surprising that we should find the golden ratio appearing in this Platonic solid.

(a)   (b)   (c)

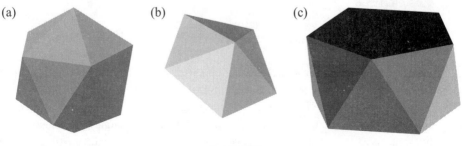

FIGURE **19.5**

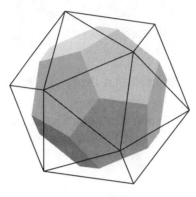

FIGURE **19.6**

Indeed, there are other pentagons within the icosahedron—line segments joining the centers of adjacent faces of an icosahedron form the edges of a dodecahedron, as illustrated in Figure 19.6.

Similarly, line segments joining the centers of adjacent faces of a dodecahedron form the edges of an icosahedron, a result we summarize by referring to the dodecahedron and icosahedron as *dual polyhedra*. We will encounter duality of polyhedra again.

We'll now show that within every icosahedron are *golden rectangles,* rectangles for which the ratio of the longer side to the shorter side is $\phi = 1.618\ldots$. This fact leads to an interesting way to construct a model of the icosahedron. First cut out three golden rectangles of dimensions $2 \times 2\phi$ from heavy cardboard, cut slits as shown in Figure 19.7(a), and fit the rectangles together as shown in Figure 19.7(b).

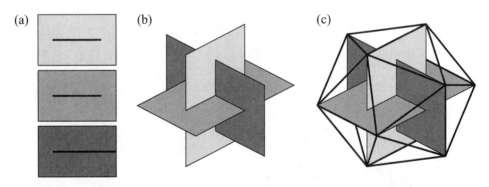

FIGURE **19.7**

Now make a small hole near each vertex, and with a piece of string or yarn connect adjacent vertices as shown in Figure 19.7(c) to obtain the edges of an icosahedron. To show that this construction works, we superimpose an xyz coordinate system on the rectangles so that the origin is at the center and each rectangle lies in one of the coordinate planes. The coordinates of the twelve vertices of the rectangles are $(0, \pm1, \pm\phi)$, $(\pm\phi, 0, \pm1)$, and $(\pm1, \pm\phi, 0)$. Then the distance d between adjacent vertices is (recalling

that $\phi^2 = \phi + 1$)

$$d = \sqrt{1^2 + (\phi - 1)^2 + \phi^2} = \sqrt{4} = 2,$$

exactly the same as the length of the shorter side of each rectangle.

The octahedron's secret: a structure for a table

The octahedron is generally pictured with one vertex at the bottom and the opposite vertex at the top, i.e., we view the solid as two square pyramids joined at a common base, as illustrated in Figure 19.8(a).

(a)   (b)

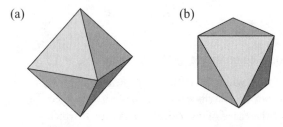

FIGURE **19.8**

However, if we place the octahedron on one of its triangular faces, then another parallel face rotated 180° appears on the top, and the remaining six faces appear joining the top vertices with the corresponding lower vertices, as shown in Figure 19.8(b) In this orientation we see that the octagon is also a triangular antiprism. The edges of the octahedron make a great structure for a table: the three vertices on the floor give stability to the table; while the three vertices at the top can support another object. Figure 19.9 shows two industrial examples. Fitted with a circular glass top, tables with this design could also be used as home furniture.

FIGURE **19.9**

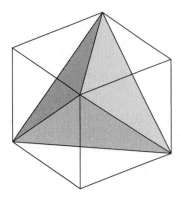

FIGURE **19.10**

The cube's secret: a union of triangular pyramids

The cube has six faces, while the tetrahedron has six edges, so it is not surprising to learn that a tetrahedron can be inscribed in a cube so that each edge of the tetrahedron is the diagonal of one of the faces of the cube. See Figure 19.10.

Outside each face of the tetrahedron lie three edges and a vertex of the cube, which with the face of the tetrahedron form a triangular pyramid with isosceles right triangles for three of its four faces. Only three of these pyramids can be seen in the drawing on the left in Figure 19.10, the fourth is in the rear. On the right in Figure 19.10 is a photograph of a sculpture on the campus of Hampshire College in Amherst, Massachusetts, in which the tetrahedron and all four triangular pyramids are visible. Thus every cube is the union of five triangular pyramids—a central tetrahedron plus four others on the faces of the tetrahedron.

Line segments joining the centers of the adjacent faces of a cube form the edges of an octahedron and conversely, hence the cube and the octahedron are dual polyhedra like the dodecahedron and the icosahedron.

The tetrahedron's secret: a better kite

Take another look at Figure 19.1. The tetrahedron is different from the other four Platonic solids in that it doesn't have pairs of parallel faces, pairs of parallel edges, or pairs of vertices that lie on opposite sides of the center of the polygon. The tetrahedron also differs from the other Platonic solids in that it is *self-dual*, that is, line segments joining the centers of the adjacent faces of a tetrahedron form the edges of another tetrahedron. However, it is not the only self-dual polyhedron—pyramids are also self-dual. Finally, general tetrahedra (with general triangles as faces) are the only convex polyhedra with four faces.

Alexander Graham Bell (1847–1922), the inventor of the telephone, was an advocate of use of the tetrahedron in the design of kites [Bell, 1903]. Bell realized that the rigidity of the tetrahedral structure was superior to the cubical structure of box kites, and that many-celled tetrahedral kites could be built from the basic four-cell structure illustrated in Figure 19.11. Plans for constructing tetrahedral kites can be found easily on the web.

(a)   (b)

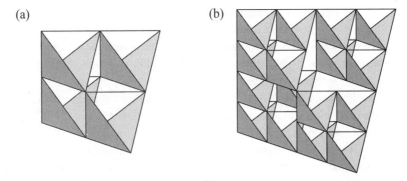

FIGURE **19.11**

As Bell wrote in 1903, "The tetrahedral principle enables us to construct out of light materials solid frameworks of almost any desired form, and the resulting structures are admirably adapted for the support of aero-surfaces of any desired kind, size, or shape. Of course the use of a tetrahedral cell is not limited to the construction of a framework for kites and flying-machines. It is applicable to any kind of structure whatever in which it is desirable to combine the qualities of strength and lightness. Just as we can build houses of all kinds out of bricks, so we can build structures of all sorts out of tetrahedral frames, and the structures can be so formed as to possess the same qualities of strength and lightness which are characteristic of the individual cells. I have already built a house, a framework for a giant wind-break, three or four boats, as well as several forms of kites, out of these elements."

Polyhedral dice

An age-old application of the Platonic solids is in the design of dice. Cubical dice were used by the Romans for gambling and in a game similar to backgammon, and remain today the most familiar shape for dice. Of course, other polyhedra can be and are used for dice, especial in role-playing games, war games, and for fortune telling. In Figure 19.12, we see the five Platonic dice and one more.

FIGURE **19.12**

Each of the Platonic solids except the tetrahedron has a pair of parallel faces, so that there is always an "upper" face when it is rolled as a die. The tetrahedron lands with a vertex pointing upwards, and so tetrahedral dice are designed with three numbers on each face. The number rolled is then taken as the number at the bottom (a 4 in the case of the tetrahedral die on the right in Figure 19.12).

Since Platonic dice have 4, 6, 8, 12, and 20 sides, they are "random number generators" for the uniform distribution on n numbers for $n = 4, 6, 8, 12,$ and 20. The polyhedral die in the upper left of Figure 19.12 (with faces 6, 2, and 8 visible) is a *pentagonal trapezohedron,* a solid with ten congruent kite-shaped faces (it is the dual of the pentagonal antiprism illustrated in Figure 19.5(c)). Dice with this shape are popular with teachers of probability and statistics, since they enable students to simulate random decimal digits. Of course, one could also use the icosahedron for this purpose, using each of the digits $\{0, 1, 2, \ldots, 9\}$ twice as a face.

19.2   The rhombic dodecahedron

The cube is the unique *space-filling* Platonic solid (a space-filling polyhedron is a polyhedron which "tiles" space, analogous to the way certain polygons tiled the plane in Chapter 10). Another space-filling solid is the *rhombic dodecahedron,* a solid with twelve rhombi as faces. In Figure 19.13, we see (a) a picture of the rhombic dodecahedron, (b) the shape of each face, and (c) a plan which can be used to construct one.

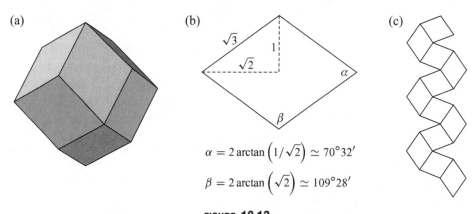

(a)   (b)   (c)

$$\alpha = 2\arctan\left(1/\sqrt{2}\right) \simeq 70°32'$$

$$\beta = 2\arctan\left(\sqrt{2}\right) \simeq 109°28'$$

FIGURE **19.13**

The following construction, adapted from [Senechal and Fleck, 1988], shows that rhombic dodecahedra "pack space," i.e., are space-filling solids.

(i) Construct a regular octahedron with side length 14.1 cm ($14.1 \simeq 10\sqrt{2}$).

(ii) Construct a cube with side length 10 cm.

(iii) Construct six pyramids each with a square base 10 cm by 10 cm, and isosceles triangular faces 10 cm by 8.7 cm by 8.7 cm ($8.7 \simeq 5\sqrt{3}$).

(iv) Construct eight pyramids each with equilateral triangular base 14.1 cm on a side and isosceles triangular faces 14.1 cm by 8.7 cm by 8.7 cm.

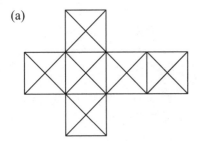
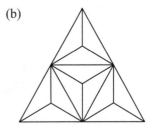

(a)

(b)

FIGURE 19.14

(v) Arrange the six pyramids from step (iii) in the shape of a cross, as shown in Figure 19.14(a). When the pyramids are folded inwards, they form a cube, with the six pyramid apexes meeting at its center. When the six pyramids are glued to the faces of the cube from step (i), we obtain a rhombic dodecahedron. The volume of the rhombic dodecahedron is thus twice the volume of the cube; and edges of the cube correspond to the short diagonals of the rhombic faces. Thus rhombic dodecahedra pack space, since cubes do.

(vi) Similarly arrange four of the triangular pyramids from step (iv) as shown in Figure 19.14(b). When folded inwards, they form a tetrahedron.

(vii) When the eight triangular pyramids are glued to the faces of the octahedron from step (i), they form (again!) a rhombic dodecahedron, congruent to the one constructed in step (v). The edges of the octahedron form the long diagonals of the rhombic faces. Since rhombic dodecahedra pack space, a collection of tetrahedra and octahedra must as well (although neither does alone).

19.3   The Fermat point again

In Section 6.4 we encountered the concept of the Fermat point of a triangle—the point F in the interior of an acute triangle ABC such that the sum $\overline{FA} + \overline{FB} + \overline{FC}$ of its distances to the vertices is minimal. To find F, we constructed equilateral triangles on the sides of $\triangle ABC$ and connected each vertex of $\triangle ABC$ to the exterior vertex of the equilateral triangle on the opposite side. The Fermat point F is the intersection of these three lines.

One can visualize the Fermat point of a triangle by means of soap films. Prepare two sheets of plastic, and place three upright pegs between them at the vertices of $\triangle ABC$. Holding the structure by one of its corners, dip it into the soap solution. Three soap films will appear between the pegs, intersecting at the Fermat point, as illustrated in Figure 19.15.

The surface tension in the soap film minimizes the surface area of the soap film, pulling it into a stable position. In our case, the sum of the distances of the point of intersection to the vertices (the pegs) will be minimal.

The same idea can be used to determine Fermat points for any given set of points. What do you think happens with four points? Although this problem can be solved analytically for specific sets of four points, the general case is unsolved.

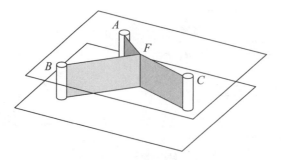

FIGURE **19.15**

Pedagogical remark. Joining the Fermat point to any two vertices forms a 120° angle (recall Challenge 6.4). This fact gives us another method of locating the Fermat point. Prepare two transparencies, one with the triangle ABC, the second with three line segments emanating from a central point F at 120° angles. When the second transparency is placed on top of the first so that the line segments pass through the vertices of the triangle, F will be at the Fermat point.

19.4 Challenges

19.1 How many different (up to symmetry) hexominos can be formed from six congruent squares (where each square has at least an edge in common with another square)? Which hexominos will correspond to the unfolding of a cube?

19.2 In a standard cubical die, the numbers of spots on opposite faces sum to seven. As a consequence, how many ways can one construct a die?

19.3 Given any natural number $n > 0$, show how to construct a die with the digits $1, 2, \ldots, n$ with equal probabilities.

19.4 Can a rhombic dodecahedron be used as a fair die?

19.5 Which regular polygons can be obtained as planar sections of a cube? of a tetrahedron? of an octahedron?

19.6 In Figure 19.16 we have a collection of tori. What planar curves are possible sections of a torus?

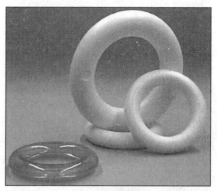

FIGURE **19.16**

19.7 A rhombic dodecahedron (Section 19.2) results from adding six pyramids to a cube of side length 1. What is the side length of the rhombic faces?

19.8 A *gyrobifastigium* is a solid formed by joining two triangular prisms at their square bases, as show in Figure 19.17. Show that the gyrobifastigium also packs space. [Hint: Make a model.]

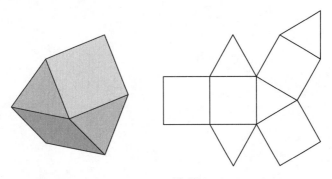

FIGURE **19.17**

20

Combining Techniques

In solving mathematical problems, it is often advantageous to combine various problem-solving techniques. The same is true for creating visual proofs of mathematical theorems. In this chapter we present a variety of examples combining many of the techniques found in earlier chapters.

20.1 Heron's formula

Heron's remarkable formula $K = \sqrt{s(s-a)(s-b)(s-c)}$ for the area K of a triangle with side lengths a, b, and c, and *semiperimeter* $s = (a+b+c)/2$ can be proven by a variety of methods (see [Nelsen, 2001] for references). In this section we present visual proofs of two lemmas that reduce the proof of Heron's formula to elementary algebra.

Let $\triangle ABC$ be a triangle with sides a, b, c, as in Figure 20.1(a), and bisect each angle to locate the center of the *incircle* (as did Heron). Extending an *inradius* (length r) to each side now partitions the triangle into six smaller right triangles, with side lengths as indicated in Figure 20.1(b).

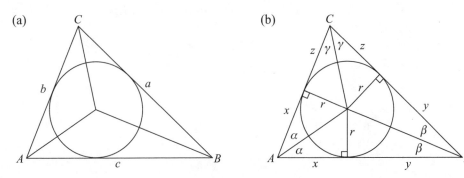

FIGURE 20.1

Lemma 1. *The area K of a triangle is equal to the product of its inradius and semiperimeter.*

Proof. Since the six interior triangles are all right triangles, they can be arranged into a rectangle whose dimensions are r by $x + y + z = s$, as shown in Figure 20.2.

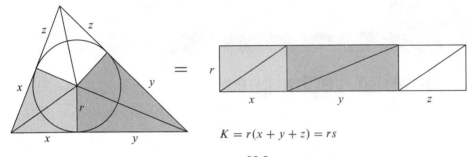

$$K = r(x + y + z) = rs$$

FIGURE **20.2**

Lemma 2. *If α, β, and γ are any positive angles such that $\alpha + \beta + \gamma = \pi/2$, then*

$$\tan\alpha \tan\beta + \tan\beta \tan\gamma + \tan\gamma \tan\alpha = 1.$$

Proof. First construct a right triangle with acute angle α and legs of lengths 1 and $\tan\alpha$ as shown in Figure 20.3. Next construct the right triangle with acute angle β, with legs as shown in Figure 20.3; and then the smaller right triangle with acute angle α, and legs of lengths $\tan\beta$ and $\tan\alpha \tan\beta$. Finally add the right triangle with acute angle γ. Since $\alpha + \beta + \gamma = \pi/2$, the resulting figure is a rectangle, and the vertical sides have the same lengths, yielding the desired result.

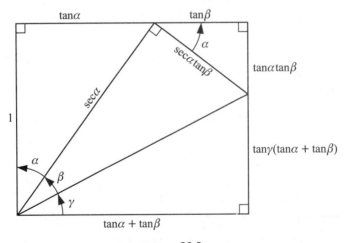

FIGURE **20.3**

Theorem (Heron's Formula). *The area K of a triangle with sides of length a, b, c and semiperimeter* $s = (a + b + c)/2$ *is given by*

$$K = \sqrt{s(s-a)(s-b)(s-c)}.$$

Proof. Applying Lemma 2 to the angles α, β, and γ in Figure 20.1(b) yields

$$1 = \tan\alpha\tan\beta + \tan\beta\tan\gamma + \tan\gamma\tan\alpha$$

$$= \frac{r}{x}\cdot\frac{r}{y} + \frac{r}{y}\cdot\frac{r}{z} + \frac{r}{z}\cdot\frac{r}{x}$$

$$= \frac{r^2(x+y+z)}{xyz} = \frac{r^2 s}{xyz} = \frac{K^2}{sxyz}$$

where the last step follows from Lemma 1. Since the semiperimeter s satisfies $s = x + y + z = x + a = y + b = z + c$, it follows that

$$K^2 = sxyz = s(s-a)(s-b)(s-c),$$

which completes the proof.

20.2   The quadrilateral law

In Section 12.1, we saw that for any parallelogram, the sum of the squares on the diagonals is equal to the sum of the squares on the sides, a result we called the parallelogram law. We now generalize that result to arbitrary convex quadrilaterals.

Theorem. *In any convex quadrilateral the sum of the squares of the sides is equal to the sum of the squares of the diagonals plus four times the square of the distance between the midpoints of the diagonals. See Figure 20.4.*

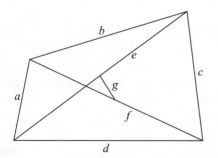

$$a^2 + b^2 + c^2 + d^2 = e^2 + f^2 + 4g^2$$

FIGURE 20.4

Proof. Joining the midpoints of the sides of the quadrilateral to the midpoints of the diagonals, as shown in Figure 20.5(a), creates two parallelograms with side lengths as labeled (recall that a line segment joining midpoints of two sides of a triangle is parallel to and one-half the length of the third side).

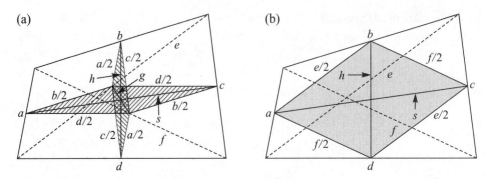

FIGURE **20.5**

Applying the parallelogram law to the shaded parallelograms in Figure 20.5(a) yields

$$g^2 + h^2 = 2\left(\frac{a}{2}\right)^2 + 2\left(\frac{c}{2}\right)^2 \quad \text{and} \quad g^2 + s^2 = 2\left(\frac{b}{2}\right)^2 + 2\left(\frac{d}{2}\right)^2.$$

Adding, we have

$$2g^2 + s^2 + h^2 = 2\left(\frac{a}{2}\right)^2 + 2\left(\frac{b}{2}\right)^2 + 2\left(\frac{c}{2}\right)^2 + 2\left(\frac{d}{2}\right)^2.$$

Connecting the midpoints of the sides of the quadrilateral as shown in Figure 20.5(b) yields another parallelogram, so that

$$s^2 + h^2 = 2\left(\frac{e}{2}\right)^2 + 2\left(\frac{f}{2}\right)^2.$$

Combining the last two displays (and clearing fractions) yields the desired result.

20.3   Ptolemy's inequality

In Section 7.1 we encountered Ptolemy's theorem: In a quadrilateral inscribed in a circle, the product of the length of the diagonals is equal to the sum of the products of the lengths of the opposite sides. But what happens with a general quadrilateral? In the case of a general *convex* quadrilateral, the product of the length of the diagonals is *less than or equal* to the sum of the products of the lengths of the opposite sides, as the sequence of frames in Figure 20.6 [Alsina, 2005] illustrates.

The first frame exhibits the quadrilateral. In the second frame, the side with length b has been rotated to coincide with the diagonal of length e, and a segment drawn parallel to the side of length d. This creates a triangle (in gray) similar to the triangle with sides of lengths c, d, and e.

In the third frame we rotate the gray triangle to the position shown, and draw the segment of length x. Note that $bd/e + x \geq f$. The striped triangle is similar to the triangle with sides a, b, and e (note that the sides adjacent to the common marked angle are proportional), and hence it can be translated to the position shown in the fourth frame. It now follows that $x = ac/e$, so that $bd/e + ac/e \geq f$, or equivalently, $ef \leq ac + bd$.

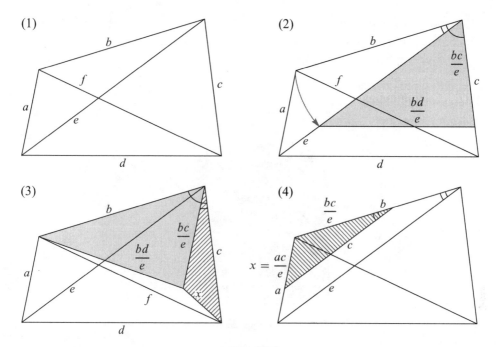

FIGURE **20.6**

20.4 Another minimal path

In Section 9.4 we saw how to find the shortest path between a spider and a fly in a rectangular room; and in Challenge 6 in Section 18.1 we found the shortest path between two points on the surface of a cylinder. What about minimal paths on the surface of a cone? Consider this related problem [Steinhaus, 1938]: Suppose we are at a point on the cone (not the vertex), and wish to travel around the cone, as shown in Figure 20.7(a). What path has minimal length?

Here we proceed as before, by converting a three-dimensional problem to a two-dimensional one. To do so, we cut the cone along the dotted line in Figure 20.7(a), which is directly opposite the line from the vertex through the given point, and open the

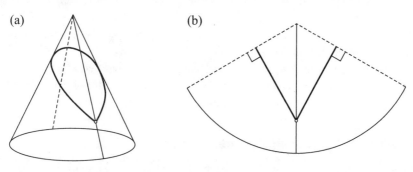

FIGURE **20.7**

cone as shown in Figure 20.7(b). The surface of the cone is now a sector of a circle, and the shortest path consists of the two lines perpendicular to the lines resulting from the cut.

However, for some cones the solution is different—see Challenge 20.7.

20.5 Slicing cubes

FIGURE **20.8**

Suppose we have a large cube such as the one in Figure 20.8, and we wish to cut it into a large number, say one hundred, or even one million, smaller cubes. If we are permitted to rearrange and stack pieces between cuts, what is the minimum number of cuts required?

Here's an opportunity to experiment with "smaller" problems. Let $f(n)$ denote the minimum number of cuts required to produce n^3 smaller cubes. Trivially $f(1) = 0$ and $f(2) = 3$. To produce 27 small cubes, six cuts will be required since the central small cube has six faces, no two of which can be obtained by any single cut. Thus $f(3) = 6$. Rearranging stacks between cuts shows that $f(4) = 6$, and $f(5) = f(6) = f(7) = f(8) = 9$.

For higher values of n we can continue to experiment with cubes, or consider a lower-dimensional problem. Suppose we wish to cut a square of cardboard into n^2 smaller squares, or that we wish to cut a rod into n pieces, again permitting rearrangement and stacking between cuts. The rod problem is probably the easiest to visualize, and it is easy to see that when n lies between successive powers of two, i.e., $2^{k-1} < n \leq 2^k$, k cuts are required, so that to obtain n pieces we need $\lceil \log_2 n \rceil$ cuts, where $\lceil x \rceil$ denotes the least integer greater than or equal to x. Since we need $\lceil \log_2 n \rceil$ in each dimension, we suspect that for the cube $f(n) = 3\lceil \log_2 n \rceil$ cuts are required. See [Tanton, 2001a] for an elegant induction proof of this result. Thus to obtain one million small cubes, only $f(100) = 3\lceil \log_2 100 \rceil = 21$ cuts are required.

20.6 Vertices, faces, and polyhedra

Consider the following question: if two polyhedra have exactly the same points in space as vertices, are they congruent? The answer is yes if we are dealing with convex polyhedra.

(a)   (b)

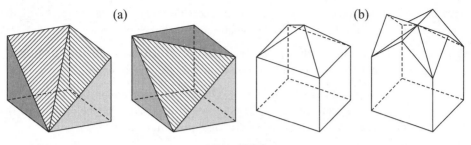

FIGURE **20.9**

But if we permit concave polyhedra, the answer is no, as the two polyhedra in Figure 20.9(a) illustrate. Both have the same seven vertices, but the one on the left is concave while the one on the right is convex.

A related question is the following: if two polyhedra have all their faces in the same planes, are they congruent? Figure 20.9(b) shows a convex polyhedron and a concave polyhedron such that identical planes determine the faces.

Using three-dimensional models—either real or virtual—can help in finding counterexamples in spatial situations where results may or may not be the same as the corresponding results in the plane.

20.7 Challenges

20.1 Is there a quadrilateral law for non-convex quadrilaterals, such as those shown below?

FIGURE **20.10**

[Hint: Experiment with pictures analogous to those in Figure 20.4.]

20.2 Does Ptolemy's inequality hold for quadrilaterals such as those in Challenge 20.1?

20.3 If α, β, and γ are angles in an acute triangle, show that

$$\tan \alpha + \tan \beta + \tan \gamma = \tan \alpha \tan \beta \tan \gamma.$$

[Hint: a picture similar to Figure 20.3 can be used.]

20.4 Given $a, b > 0$, find the extrema of the function $f(t) = a \cos t + b \sin t$.

20.5 Suppose that a, b, c are the side lengths of a given triangle. Construct a triangle whose medians have lengths a, b, c.

20.6 Show how to construct a triangle given the side length a, the length h_a of the altitude to a, and the length h_b of the altitude to another side b.

20.7 Show that if the base radius r and the height h of the cone in Section 20.4 satisfy $h \geq r\sqrt{3}$, then there is no path around the cone of minimal length.

20.8 Consider the planar versions of the two questions from Section 20.6: (a) If two polygons have the same points in the plane as vertices, must they be congruent? (b) If two polygons have their edges in the same lines, must they be congruent?

20.9 This is an open problem in visualization: Given a convex quadrilateral with side lengths a, b, c, d, semiperimeter $s = (a+b+c+d)/2$, and which is inscribed in a circle, find a visual proof of *Brahmagupta's formula* for the area of the quadrilateral, $K = \sqrt{(s-a)(s-b)(s-c)(s-d)}$.

Part II

Visualization in the Classroom

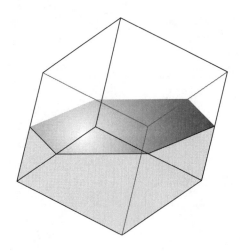 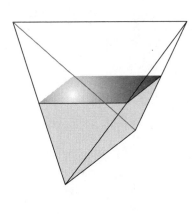

Mathematical drawings: a short historical perspective

"Geometry is to the plastic arts what grammar is to the art of the writer."
—Guillaume Apollinaire

Since its very beginning, the discipline of mathematics has combined three different resources for its development: a *natural language* (hieroglyphic Egyptian, Greek, Latin, English,...), a *symbolic language* (with signs $+$, $-$, $\times$, $=$,... and symbols x, y, z, f,...) and *figures*. There were two key reasons for introducing images in mathematical texts: to substitute appropriate drawings for long linguistic descriptions; and to facilitate mental reasoning based upon graphical intuition.

In this photograph of a portion of the Moscow papyrus (Egypt, circa 1850 BC) we can see text, symbols and ... the image of a trapezoid.

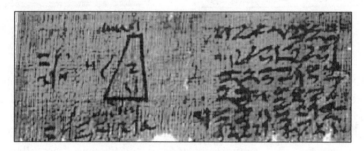

FIGURE **II.1**

In the beginning, most mathematical imagery was devoted *to illustrating* the discourse. Only in a few historical documents can one find an advanced level of drawing: *the direct visualization of mathematical properties*.

The Japanese drawing on the left in Figure II.2 shows a method of partitioning a circle to obtain a figure similar to a rectangle, and on the right the generation of a spiral.

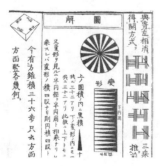 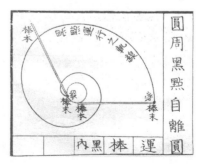

FIGURE **II.2**

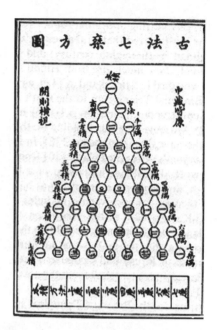

FIGURE **II.3**

In the Chinese woodcuts in Figure II.3, we see some geometrical figures superimposed on an illustration of mountains and a river; and an image of Pascal's triangle.

Great artists like Albrecht Dürer (1471–1528) introduced new styles of representation. On the left in Figure II.4 we have Dürer's drawing of an icosahedron, which includes a plane representation of the twenty triangular faces, and two wire-frame illustrations.

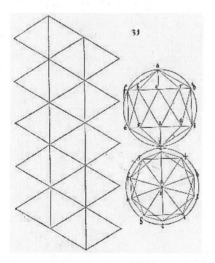
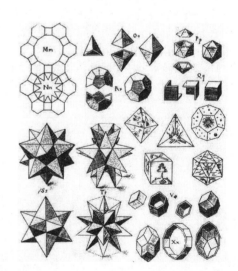

FIGURE **II.4**

FIGURE II.5

The desire to make quality pictures also motivated the development of several branches of geometry. Descriptive geometry provided the rigorous construction of perspectives. The drawing on the right in Figure II.4 is due to Johannes Kepler (1571–1630).

The invention of the printing press created new opportunities for reproducing images, and of course the most remarkable consequence was the wide distribution of books. Figure II.5 shows the title page of Hans Lencker's 1622 book entitled *Perspectiva*. For many years geometrical objects were the essential models for illustrating the use of perspective methods. This in turn further motivated the study of polyhedra.

The technological devices and software available today, such as computers and calculators with their interactive possibilities, have fostered the development of virtual images and new branches of mathematics such as computational geometry, which deals in part with challenging problems of representation by means of technical resources.

On visual thinking

> *"I have decided to introduce not a single figure in the text."*
> —Jean Dieudonné

> *"Draw a figure."*
> —George Pólya

The quotations above represent two extreme positions concerning the use of figures. For Dieudonné, as well as for many other mathematicians, the only correct way to present mathematics is by means of formal discourse based on formal languages. For Pólya, mathematical problem solving is often best done beginning with a visual representation.

In addition to the classical arguments of mathematicians to minimize the role of intuition in order to save rigor, some other objections have been raised in terms of "perception" problems: images may give a false perception of realities. For example, a recent announcement of the FlatronTM company, which manufactures flat television and computer monitors, included the claim: "Perceptions can often alter reality". This is a point of view that we will not pursue here.

For either learning mathematics or for doing mathematical research, it is clear that attention must be devoted to the development of *visual thinking*. In the heuristic of mathematical discovery, internal visualization plays a major role and, in some cases, this process may be the keystone of new research. As J. E. Littlewood wrote in his celebrated *A Mathematician's Miscellany* [Littlewood, 1953], "a heavy warning used to be given that pictures are not rigorous; this has never had its bluff called and has permanently frightened its victims." Our emphasis here will be not just on the minds of mathematicians but on the importance of visual thinking for all of us.

According to Rudolf Arnheim [Arnheim, 1970], visual thinking is "an active exploration, selection, grasping of essentials, simplification, abstraction, analysis and synthesis, completion, correction, comparison, problem solving, as well as combining, separating, putting in context,...", i.e., visual thinking is a powerful tool which can be applied in many situations.

In [Senechal, 1991] we find this definition of *visualization* and its relationship to the general framework of visual thinking: *visualization is any process producing images (pictures, objects, graphs, diagrams,...) in the service of developing visual thinking.*

According to P. R. Richard [Richard, 2004] visualization is important for *figurative inference*:

> On the one hand, it is well known that reasoning in geometry is not only based on words or on symbols, but also on drawings and visual images (mental pictures). In a situation of validation, reasoning controls the action (construction of lines or handling of drawing tools) as much as the reflection (development of visual images and of their relationship to the ideal). On the other hand, even when one admits a distinction between verbal (linguistic–symbolic) and nonverbal (non-linguistic–symbolic) reasoning, one doesn't generally consider the possibility that the latter could be associated with the forms of reasoning other than in intuitive thought or in thought that turns exclusively toward action.
>
> To bring about or to justify certain steps in reasoning, the student incorporates a drawing—or a comic strip—in the discursive structure of the proof for the same purpose as a verbal proposition, by addition of graphic reasoning to the discursive reasoning. The identification of a figurative inference occurs in as much as it is impossible to capture the reasoning locally while concealing the drawing, even if one attempts simultaneously to bring the inferred verbal proposition closer to the thematic continuity developed in proofs.

Part I of this book was devoted to the presentation of some visualization methods in the hope that interesting mathematical images will help promote ingenuity and creativity in the visual thinking process.

Visualization in the classroom

"Voir, c'est croire!"
— French proverb

Among the standards presented by the National Council of Teachers of Mathematics in 2000 [NCTM, 2000] we find problem solving, reasoning, proof, communication, connections and representation. All of these require the student to think visually or may be achieved by means which incorporate visualization.

Historically, visualization in the classroom has occurred with pencil on paper or with chalk on the blackboard. While this practice may be changing with more and more students having access to computers and graphing calculators, the traditional methods may never completely disappear.

FIGURE II.6

Technology opens new possibilities to visual experiences, from the modest superposition of transparencies on an overhead projector to the latest software (Cabri II™, Cabri 3D™, Geometer's Sketchpad®, Cinderella, Mathematica®, Maple®, Derive™, Matlab®, Geup,...) which can be used to make precise drawings by computer and to project them on a screen. Internet resources also provide a large collection of high quality pictures which can be used in the classroom.

Beyond the tools used for visualization, from ordinary chalk to the latest software, visualization in the classroom has its own pedagogical values. In the *Principles and Standards for School Mathematics* [NCTM, 2000] it is clearly stated in the Representation Standard that students need to "select, apply, and translate among mathematical representations to solve problems." Today in addition to the classical modes of representation (enactive, iconic and symbolic), one may think in a wider framework of mixed representations, e.g., in [Wong, 2004] the author presents a multi-modal think-board comprising numbers, symbols, words, diagrams, real objects and stories, where under the label of "diagram" Wong includes illustrations, pictures, graphs, charts and figures.

As the outset, visualization may be a tool to develop intuition, to start solving a problem or a natural way to identify concepts. But it also deserves a central role in the important task of creating proofs. Werner Blum [Blum and Kirsch, 1991; 1996] has been working with so-called *reality-based proofs* (or contextual proofs), arguments, supported by contextualized knowledge that "justify" or "convince." For example, understanding that "if the derivative f' of a function f vanishes on an interval then the function f must be constant" may be very simple if one considers the visual image of a car and a function $f(x)$ that describes the position (on some axis) of the vehicle at time x. Then $f'(x)$ is the instantaneous velocity of the car. If $f'(x) = 0$ for some time interval (the car does not move) then $f(x)$ is constant (the car stays where it is).

Proof, for the mathematician, is an essential component of research, but proofs in the classrooms may have the added value of *explaining* the properties under consideration. Recently G. Hanna and H. N. Jahnke [Hanna and Jahnke, 2004] wrote the following:

> Clearly, an explanatory proof in school mathematics, as in any other context, must be one that not only demonstrates the truth of its assertions, but also helps one understand why the assertions are true. The aim of such a proof is always to bring to light underlying relationships that place its assertions in a broader mathematical context. In the classroom, however, an explanatory proof must rely upon the more limited mathematical knowledge of students and make use of the properties of objects best known to them. A number of different methods have been employed to this end, such as the judicial use of visualisations [Hanna, 1990], explorations with dynamic software [de Villiers, 1999], pencil-and-paper proofs appropriate to the cognitive development of the student, or the use of arguments from physics.

As an example of the use of an argument from physics, consider the following statement: If we have a convex polyhedron and we assign to its ith face (with area A_i) the unit outward pointing normal vector $\vec{n}_i$, then $\sum A_i \vec{n}_i = 0$. This property follows from a physical consideration [Prasolov, 2000]:

> Let us fill the polyhedron with gas. The pressure of the gas on the ith face is proportional to $A_i \vec{n}_i$, while the sum of all the pressure forces on the faces is zero (otherwise a *perpetuum mobile* could have been constructed).

Mathematical proofs are perhaps the most remarkable aspect of mathematicians' work. Proofs are at the heart of mathematics [Davis and Hersh, 1981], [De Villiers, 1999], [Rotman, 1998], they constitute the goal of most mathematical research and need to be formalized in their final versions. Proofs and the act of proving have also been shown to be of great pedagogical value insofar as they aid students to gain a better understanding of mathematics [Hanna, 1990], [Lakatos, 1976], [Pólya, 1954, 1981]. Therefore, the key issues are how to construct appropriate exercises involving proof in the classroom at all levels (for appropriate topics) and how to avoid unnecessary formalism and rigidity of presentation that may squelch learners' interest in mathematics.

On the role of hands-on materials

In Chapters 1–17 we focused our attention on graphical images and in Chapters 18, 19 and 20 we presented a small sample of materials that facilitate an intuitive approach to mathematics, or to develop spatial intuition. What is the value of experimentation in mathematics? Following [De Villiers, 2003] we can consider experimentation as comprising non-deductive methods including intuitive, inductive or analogical reasoning. Its important aspects are:

- *conjecturing* (looking for an inductive pattern, generalization,...;
- *verification* (obtaining with certainty the truth or validity of a statement or conjecture);
- *global refutation* (disproving a false statement by generating a counterexample);
- *heuristic refutation* (reformulating, refining or polishing a statement by means of local counterexamples);
- *understanding* (the meaning of a proposition, concept or definition or assisting with the discovery of a proof).

These considerations make it clear that experimentation is important in mathematics and that it plays a significant role in learning.

So by organizing a mathematics laboratory with hands-on materials, or by bringing them in to the classroom one can provide students with materials to help them develop visual thinking in three dimensions [Alsina, 2006].

Spatial questions and visual reasoning have received a great deal of attention in the literature of mathematics education (see, e.g., [Bosch, 1994], [Brown, 1999], [Davis, 1993], [Davis and Hersh, 1981], [Dreyfus, 1994], [Hanna and Jahnke, 1996], [Hersh, 1993], [Lakatos, 1976], [Mason, 2004], [Richard, 2004], [Senechal, 1991], and the special publications [ZDM, 1994] and [Zimmermann and Cunningham, 1991]). But in many instances teachers are not confident in dealing with three-dimensional geometry, there is a lack of good three-dimensional models in the teaching resource catalogues, and what is worse, many children end compulsory schooling without spatial literacy.

Some people believe that making models and experimenting may have a role in the early grades, but that it is something to be replaced by more sophisticated linguistic and symbolic descriptions later on, i.e., "real mathematics comes after experimental work." This belief is incorrect, as research has shown that if we do not provide a stimulating reference for abstract concepts, then formal approaches degenerate into merely an intellectual game. Visual thinking is not just an appetizer for the main course of abstraction. Clearly, at certain levels one is restricted to a selection of spatial items, but there are opportunities to offer a broader *spatial culture* at all ages.

Spatial common sense must be cultivated and developed since it is not necessarily an innate capability. This is not "spatial awareness" or a culturally dependent "average understanding." If spatial common sense is not properly cultivated and worked on in mathematical activities, then students may have difficulties in problem solving when faced with three-dimensional objects or transformations. In particular, if this sense is not developed, students may well try to apply sophisticated, unnatural solutions to spatial questions that often have quite obvious answers.

There is a general feeling among some teachers that "mathematical difficulties necessarily require a particular curricular ordering of concepts." This implies, for example, not dealing with spatial figures until study of the plane has been completed, not presenting certain shapes because their equations are of a higher degree, not presenting surfaces before having studied curves, etc. The result is generally very poor familiarity with the mathematics of three-dimensional space. By finding "appropriate descriptions" one can explore space without using lack of technical difficulty as a criterion for selection. Intuition may allow one to encounter aspects of reality that the majority of citizens would never be able to see if advanced mathematical knowledge was required.

While our main focus is on the understanding of space, it is also true that an exploration of space provides opportunities to exhibit links to the history of mankind's knowledge of space, artistic considerations (sculpture, perspective, design,...), architectural solutions (buildings, structures,...), the impact of technology (robots, machines, virtual reality,...), natural resources (deserts, forests, rivers, mountains,...), etc. For such applications see, e.g., [Bolt, 1991], [Cook, 1979], [Senechal, 1991], [Malkevitch, 1991], [Steen, 1994], among others.

While the plane is rich with mathematical concepts, space offers intriguing open-ended questions that can be addressed in a spirit of research. We should promote the idea that mathematics is a living field with many interesting problems to be posed and/or solved. Three-dimensional space is an excellent place to begin.

To sum up, the following are four important contributions that hands-on materials make in school classrooms [Alsina, 2005]:

> *Hands-on materials may open windows to creative solutions that are impossible using traditional tools.*

Whether a problem is solvable or not often depends on the resources at one's disposal. Consider, for example, traditional geometrical problems that make use of compass and straightedge. The Greek tradition of using only these two tools is quite restrictive, and while historically and mathematically important those tools are insufficient to reach the solution of many elementary problems. If one adds more versatile tools, then the situation changes completely. For example, if a ruled straightedge or similar tools are allowed for making measurements, then one can trisect any angle; if one complements the straightedge and compass with a rotating cylinder and pencils, then one can mark lines 2π units long so that squaring the circle is possible; if one has perfect graphs of curves, then the duplication of the cube is immediate, and so on.

> *Images and hands-on materials may be needed if the problem in question requires an explicit practical solution.*

When one considers mathematical problems such as "make a perspective drawing of an object whose projections are..." or "find the possible regular polygons that are planar sections of a cube," the solutions being sought include graphical elements. Often the most reasonable way to represent a cube is not as a collection of vertex coordinates and the equations of the faces, but rather as a three-dimensional object.

> *Hands-on materials may facilitate visual thinking, and constitute a more important step than making plane representations or more formal calculations.*

Of course direct observations about the reality around us or well-prepared models aid students in developing intuitive thinking skills and in discovering relationships that afterwards may be treated in a more formal way.

> *Images and hands-on materials may be the only feasible way to exhibit examples*
> *of or solutions to planar or spatial problems.*

In dissection problems involving planar or spatial figures either one gives an explicit solution or demonstrates that our solution exists. As an example, when working with tilings, dissections or with *n*-ominoes, (recall Chapters 10, 13 and 16) one deals with transformations in which areas are preserved but perimeters may change. In the three-dimensional realm, the need for models is even more important. Models are helpful but maybe not necessary for these two-dimensional problems. If one were dealing with three-dimensional tilings or *n*-ominoes, models would be essential.

Everyday life objects as resources

To study the geometric characteristics of a building you could make a scale model of it ... but there is a better alternative: use the real building! In this section we invite you to use materials readily available as objects with mathematical interest.

Most of the man-made shapes that we see around us are the result of design: houses, streets, cars, beds, bells, pencils, etc. In this designed reality there is a strong mathematical component, from sizes to shapes. Most of these objects have been designed to satisfy some desirable purpose or function. In the classical dialogue between form and function, designers look for "optimal solutions." But "optimal" may hide different ideas: minimal quantity of materials, low cost, ecological considerations, transportability goals, etc. We are interested in making geometrical shapes visible by means of daily life objects.

An interesting pedagogical exercise is to consider an everyday object and to discover all mathematical aspects associated with the object. For example consider an umbrella. This object is a flexible octagonal pyramid with several articulated parallelograms inside to hold the structure open. How do angles change in the opening or closing of the umbrella?

Beyond objects, there are daily life situations that are mathematically rich, e.g., ice cream comes in rectangular or cylindrical packages, from which we make spherical ice cream balls that we place in cones, ...

The home is also a realm of mathematics: what are the relationships among the various measures one finds for the roof of a house (with ridges, valleys, dormer windows, etc.). How many helices can you identify in your home?

It's interesting to learn how designers work and how they find solutions to design problems. Consider the story of Jacob Rabinow (1910–1999). He worked in New York, had 229 patents for various types of inventions and when he retired wrote a book entitled *Inventing for Fun and Profit* [Rabinow, 1990]. One of Rabinow's favorite topics was screws and screwdrivers. He wanted to solve the problem where screws were often removed from objects in public places because one or two screwdrivers can be used for screws of different sizes and shapes (even coins suffice to remove many screws). So Rabinow took advantage of geometry and created a screw with a head making it impossible for any conventional screwdriver to remove the screw. Here we reproduce his description:

> If you make a triangular depression with sides in the shape of three arcs, where each vertex of the triangle is the center of curvature of the opposite arc, you have a triangular hole that can be driven with a specially shaped screwdriver, but not by any flat screwdriver. If you insert a flat blade, the blade will pivot at each corner and slide over the opposite curved surface, hit the next corner and slide again, and so on. Such a screw should look very attractive and would be very difficult to open without the proper tool.

The three arcs mentioned above form the *Reuleaux triangle*, which is a convex figure of constant width that is not a circle.

We mention here how some basic geometrical shapes can be illustrated by means of common objects.

Polyhedra

Nature exhibits a fairly restrictive collection of polyhedra. Only in some specific classes of minerals can one find basic shapes like cubes or prisms. But designers have produced a wide range of objects with polyhedral form. Packaging and aesthetics have motivated many of these designs. In a geometry laboratory we can collect everyday life polyhedra (or illustrations of them). The following lists are merely suggestive of objects you will be able to find.

Polyhedra and everyday life objects

Cubes	Standard dice, bouillon cubes, Rubik's cube, boxes, etc.
Tetrahedra	3D puzzles, tripods, tetrahedral dice
Octahedra	Mineral crystals, octahedral dice
Dodecahedra	Paperweight, desk calendar, dodecahedral dice
Icosahedra	MAA logo, icosahedral dice, domes
Prisms	Toblerone™ package, candy boxes, pencils
Pyramids	Egyptian pyramids, top of an obelisk, Swarovski® crystal, plumb bob
Bipyramids	Toy tops, jewels
Other polyhedra	Jewels and jewelry, soccer balls, puzzles

Polygons

Polygons appear as faces of polyhedra, but they are used frequently in design as well.

Polygons and every day life objects

Triangles	Traffic yield sign, danger sign, musical instrument
Quadrilaterals	Sheets of paper or cardboard, tiles
Pentagons	Chrysler logo, Department of Defense building, paper strip tied into an overhand knot
Hexagons	Tiles, plates, cross-section of a pencil, hex bolts and nuts

Octagons	Stop sign, trays, tables
n-gons	Some clock faces, foreign coins
Star polygons	Star fish, Star of David

Curves

The following table collects objects to identify some curves.

Curves and every day life objects

Circle	Rim of a plate or glass, coin, wheel, ring
Ellipse	Liquid in an inclined glass, circle viewed at an angle
Parabola	Cable on a suspension bridge
Hyperbola	Profile of some bells, six arcs at the point of a sharpened pencil
Sine curve	Snake's path, sea waves
Cycloid	Trajectory of a point on a wheel
Catenary	Power lines, hanging chain
Spiral	Grooves in a vinyl record or CD, tape in a cassette, coiled rope or garden hose

To create "living" examples of curves, a group of students armed with rope and a tape measure can play a game of making curves whose special points (center, focus, etc.) are persons: circles, ellipses, parabolas, spirals,... reinforcing in a very active way the definitions of the various curves.

Quadric surfaces

A surface in three-dimensional space defined by an algebraic equation of degree two is called a *quadric surface*. Quadric surfaces are surfaces such that all planar sections are conics (ellipses, parabolas, hyperbolas, or the degenerate cases given by a point, a line, two parallel or intersecting lines). Indeed, the conic sections are the planar sections of either a cylinder or double cone, and such conics generate in space the quadric surfaces.

The trivial quadric surfaces are: the empty set, a point, a line, a plane, two parallel planes, and two intersecting planes. The non-trivial quadric surfaces are illustrated in Figure II.7. In this section we invite you to become familiar with this important class of surfaces. Although the algebraic equations that describe the surfaces can be complicated, knowledge of them is not required to obtain a visual appreciation of the surfaces.

Circular cylinder. It's the shape of round columns, round pencils, water glasses, some kitchen containers and bottles, paper rolls, cardboard tubes, etc. Paper models are easily constructed by gluing a pair of opposite edges of a sheet of paper together.

Circular cone. It's the shape of an ice-cream cone, some glassware, road markers for highway construction, the sharpened point of a pencil, or the beam from a flashlight. Paper models are readily constructed from a circular disk by cutting out a sector and gluing the straight edges together.

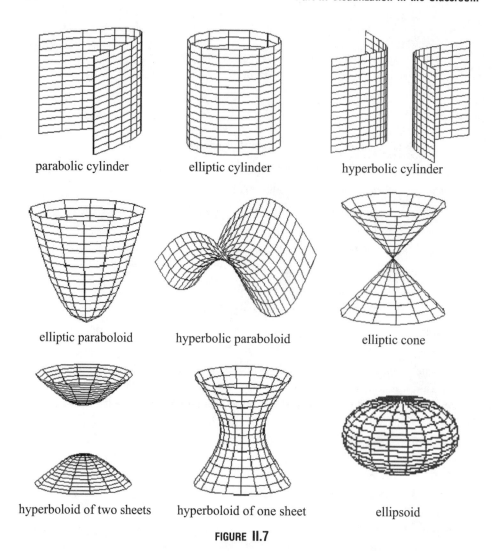

parabolic cylinder   elliptic cylinder   hyperbolic cylinder

elliptic paraboloid   hyperbolic paraboloid   elliptic cone

hyperboloid of two sheets   hyperboloid of one sheet   ellipsoid

FIGURE II.7

Conic sections of cylinders and cones. Partially fill a cylindrical or conical glass with water. When the glass is set upright on a table, the surface of the water shows a circular section. As soon as we tilt the glass we will see elliptical sections. With a conical glass, water, and a plastic cover for the glass, we can also see parabolic and hyperbolic sections. Some hourglasses are double cones and so can be used to see both branches of the hyperbola.

Ellipsoids and spheres. You'll find these shapes in many kinds of balls (rugby, soccer, ping-pong, etc.), in certain architectural domes, eggs, certain fruits, pearls, the planets, etc. With cardboard, you can make an approximate model by intersecting the three principle cross-sections. Balloons, soap bubbles, clay or putty, and the Lénárt sphere™ also make good models.

Paraboloid of revolution. Antennas used to concentrate incoming signals from satellites or light from distant stars have this shape. The headlamps of some cars have this form in order to focus the light directly ahead of the car. Some domes on buildings have a parabolic shape. If you take a glass half full of water and swirl the liquid quickly, the surface of the rotating water will be a paraboloid of revolution.

Hyperboloid of two sheets. This surface has two parts, and can be obtained by the rotation of a hyperbola (both branches) around its axis of symmetry through the foci. To construct an interesting model, cut out a cardboard disk and hang pieces of string of equal length from many points of the disk, to form a cylinder as illustrated in Figure II.8(a).

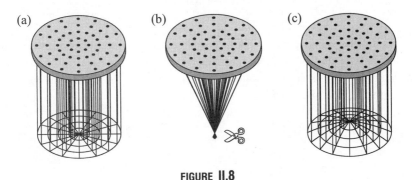

FIGURE **II.8**

Gather all the strings in a point on the central axis (you will now have a cone of strings) and cut the strings at the vertex of the cone, as in Figure II.8(b). Let the remaining strings hang freely, their end points will lie on one sheet of the hyperboloid. The reason is simple, and follows from the Pythagorean theorem, as illustrated in Figure II.9. This has an obvious implication in cutting hair at home. Be careful! The shape is also used in some telescope mirrors.

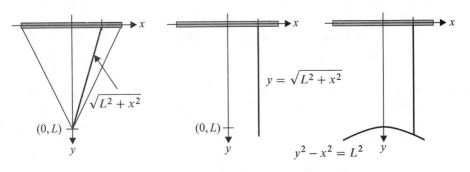

FIGURE **II.9**

Hyperboloid of one sheet. As Lord Rayleigh observed in his classic 1890 paper on bells [Rayleigh, 1964], an ideal shape for a bell is one-half of a hyperboloid of one sheet capped by a circular disk. In Figure II.10, we see one of the figures from that paper, and a photograph of the Liberty Bell in Philadelphia.

Fig. 3.

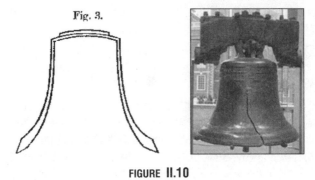

FIGURE **II.10**

To make a model of the hyperboloid of one sheet, you will need two identical parallel circular disks with uniformly spaced holes near the edge, and elastic cords stretched between corresponding holes in the disks. When you rotate one of the disks, the cords will lie on the surface of a hyperboloid of one sheet.

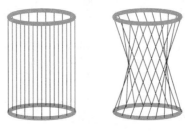

FIGURE **II.11**

Hyperbolic paraboloid. To make a model of this surface, make four identical strips of wood or plastic with uniformly spaced holes, and join them to form a quadrilateral. Stretch elastic cords between corresponding holes on opposite strips. If opposite pairs of strips are parallel, the cords lie in a plane. If the strips are not parallel, they form a skew quadrilateral, and the cords lie on a hyperbolic paraboloid. Two views are presented in Figure II.12. The architect Antoni Gaudí used this surface in his work, as we shall discuss later in this section.

So, with only simple materials, it is easy to construct visual representations of the quadric surfaces.

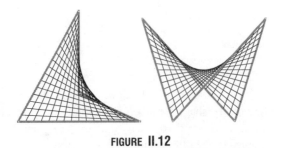

FIGURE **II.12**

Making models of polyhedra

Polyhedra are among the best known three-dimensional geometrical objects. Families of polyhedra have been studied throughout the history of mathematics and models of them have become the standard way of visualizing them. One can construct accurate models from cardboard or plastic, or purchase sophisticated models with metal or plastic edges, hinged faces, etc. There is a large market for kits of polyhedra, and polyhedra may well be the geometrical topics that receive the most attention on the Internet.

Our aim here is not to present a grand tour of the realm of polyhedra, but to present some examples and some models. Some of the important families of polyhedra, in addition to the five regular polyhedra (or Platonic solids) in Figure 19.1, are illustrated below.

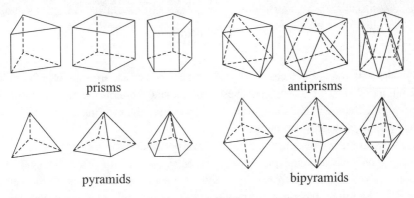

prisms   antiprisms

pyramids   bipyramids

FIGURE **II.13**

Another type of model useful for studying polyhedra are those which have transparent faces, so that interior properties may be studied. The idea is simple: we partially fill a transparent object with colored water or fine-grained sand, and tilt the object to see its various cross-sections.

Plane sections of polyhedra. For example, if we half fill a transparent cube with colored water, by tilting the cube we will see the various bisections of the cube by a plane into two parts of equal volume. In particular, we will discover that one of the cross-sections of a cube is a regular hexagon, as illustrated in Figure II.14, along with a transparent tetrahedron exhibiting a square cross-section.

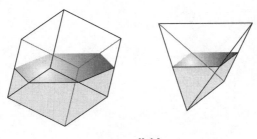

FIGURE **II.14**

Working with large models may also be of interest to students in elementary and secondary schools.

FIGURE **II.15**

Polyhedra with triangular faces are rigid, so we can construct large models of tetrahedra, octahedra, icosahedra, pyramids, etc. These can be made from plastic pipes or tubes, with flexible plastic "elbows" for the vertices. Students may then go inside the structures to measure various distances and angles.

Since there are many resources for the study of polyhedra, we can investigate polygons by focusing on the faces and planar sections of the polyhedra. While in mathematics education we often progress from polygons to polyhedra, this may not be the best way to proceed. The visual experience of students goes, in general, from three dimensions to two.

In both two and three dimensions, puzzles and tangrams constitute another interesting class of hands-on materials. On the left in Figure II.16 is a jigsaw puzzle in the shape of a heart, divided into nine pieces. When the pieces are mixed up, it can be quite a challenge to put them back together to form the original heart. Of course, other shapes can also be made from the nine pieces.

The three-dimensional puzzle on the right in Figure II.16 consists of cubes with colored paths painted on their faces, with the object of arranging the cubes to form unbroken paths of different designs.

Commercial puzzles exist in great variety to satisfy a large market for puzzle aficionados. In addition to the common jigsaw puzzles, there are many interesting geometrical

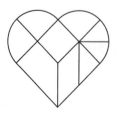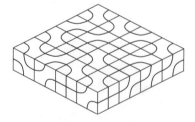

FIGURE **II.16**

puzzles consisting of two- or three-dimensional pieces. In addition to puzzles with physical manipulatives, there is a wide variety of virtual manipulative puzzles on the Internet.

For the purpose of mathematical education, we are interested in puzzles that may illustrate theorems (e.g., Pythagorean puzzles), puzzles which may help one find an ingenious solution to a three-dimensional problem, puzzles that may describe mathematical principles (e.g., measure-preserving transformations), or puzzles that may help develop spatial understanding. In elementary school, puzzles are key tools for describing figures and visualizing what they look like after rotations or flips or combinations of such operations.

We present five well-known tangrams below. Each is a decomposition of a polygon into several pieces. Here is a collection of simple exercises for use with any tangram:

(i) What kinds of polygons can be obtained using all the pieces of the tangram? How do the perimeters of these polygons differ?

(ii) What is the relationship between the line segments determined by the boundaries between the pieces of the tangram? Which pairs of line segments are parallel? perpendicular?

(iii) How are the areas of the tangram pieces related?

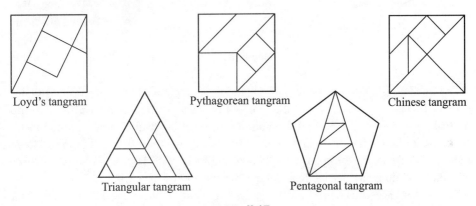

Loyd's tangram   Pythagorean tangram   Chinese tangram

Triangular tangram   Pentagonal tangram

FIGURE II.17

Of course, the same exercises can be repeated with a subset of the pieces of one of the tangrams, or with the pieces of tangrams designed by students.

Using soap bubbles

Experiments with soap solutions are both entertaining and convincing. The key observation here is the physical fact that soap films will form a minimal surface whenever they are limited by a frame.

Every child knows that when a wire ending with a circular loop is immersed into a soap solution, then blowing into the soap circle creates a soap bubble which floats in the air for a short while.

To experiment with soap films in the classroom you need to prepare your own soap solution: mix together one gallon of water, one cup of liquid detergent and one tablespoon of glycerine. Also buy florist's wire (18 gauge aluminium wire), some plastic rectangles and some pegs.

In Section 19.3 we saw how to determine the Fermat point of a triangle by means of soap films. There are many other experiments with soap films that can be used to discover other mathematical properties.

When a uniform chain is suspended from its end points, it hangs in the shape of a *catenary* curve (the function describing this curve is a modified version of the hyperbolic cosine $y = a\cosh(x/a) = a(e^{x/a} + e^{-x/a})/2$. See Figure II.18(a). When a catenary curve rotates around a line perpendicular to its axis of symmetry, it generates a surface called the *catenoid*. See Figure II.18(b).

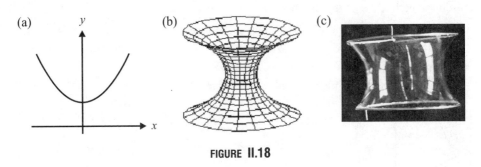

(a)   (b)   (c)

FIGURE II.18

The catenoid is a minimal surface of revolution, a property that is, indeed, exceptional. In order to have a visual model of the catenoid in space we can use a soap film. To this end, take two pieces of wire, and form similar loops at one end of each. If we immerse the loops in the soap solution and remove the loops from the water and hold the loops a short distance apart, the soap film between the loops will describe (approximately) a catenoid (the more circular the loops, the better the approximation). See Figure II.18(c).

This experience illustrates from the physical point of view the minimality of the catenoid. If you are very careful (and fast) you may be able to see that the profile of the catenoid is a catenary curve by hanging a chain or string from appropriate points behind the catenoid soap film.

If you make wire models of the regular polyhedra and immerse them in soap solution you can investigate the shapes of the soap film that will join the edges.

Lighting results

Light plays a crucial role in physics. Its maximum speed c appears in Einstein's theory of relativity ($E = mc^2$) and its behavior when reflected in a mirror (the angle of incidence is equal to the angle of reflection) or when crossing boundaries between liquids are well-known effects seen in the real world.

Let us start with a very practical problem using light. Suppose you have a table which appears to be flat. How can you be sure that it is really flat? Take an object such that you are confident about its flatness on one face (a book, a tray, a ruler,...). We just need to place this object on the table (in as many different places as we wish) and take a flashlight—if light illuminates the region between the table and the object, the table is not flat!

One of mankind's greatest accomplishments has been the visualization of time. Sand and water clocks were devised in order to mark time intervals. Wax candles, mechanical and digital devices have made it possible to indicate time in a continuous fashion. But for many centuries ingenious sundials were the most common method—associating numbers (hours) to the changing shadow of a stick or rod positioned with the appropriate inclination.

With an appropriately designed flashlight (i.e., one whose beam is conical), we can visualize the conic sections by shining the light on a wall. Which conic we obtain depends on the *direction* of the light beam relative to the wall (the direction of the cone of light is its half-axis of rotational symmetry) and the *generators* of the cone (the lines which make up the surface of the cone of light):

(i) Circle: when the direction is perpendicular to the wall;

(ii) Ellipse: when the direction intersects the wall but not perpendicularly and all generators intersect the wall;

(iii) Parabola: when the direction intersects the wall and exactly one of the generators is parallel to the wall; and

(iv) Hyperbola: in all other positions of the flashlight which illuminate the wall but different from (i), (ii), and (iii).

Note that the parabola is an exceptional shape—it is the boundary case dividing the world of ellipses and the hyperbolic world.

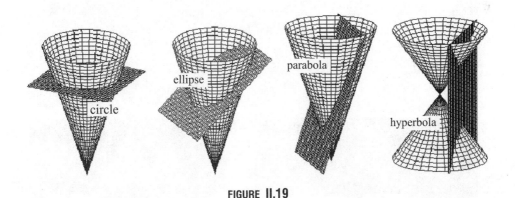

FIGURE **II.19**

With two flashlights joined at their bases and producing light in opposite directions we can visualize the two branches of a hyperbola on the wall.

Of course we may also see the conical sections by observing light from lamps in the home. We note that these experiments may even be important for those interested in theatrical lighting! Finally, one can also use light and three-dimensional models to visualize sections of polyhedra.

Mirror images

Virtual images date back to the time when prehistoric men and women first gazed into a puddle of water and saw their reflections. High quality metals provided the first technological alterative to water. These were followed by mirrors made from silvered glass. Mirrors allow us to see images obtained by reflections and provide some rather intriguing views, since reflections are isometries that change orientation.

For mirror experiments, it is possible to buy inexpensive plastic mirrors and mirrored craft paper or silvered Mylar film that can be easily cut with scissors.

With a single mirror we see reflected images in which left and right are exchanged. If we look between a pair of parallel mirrors we will see an "infinite" sequence of images in both directions of any object located between the mirrors, i.e., friezes.

With two mirrors sharing a common edge and forming an angle like an open book we can see the *rotation effect*. When we focus one eye in the direction of the intersection of the mirrors we see images of objects situated between the mirrors repeated in a circle n times when the angle between the mirrors is $360/n°$. This is the principle of the kaleidoscope. Positioning the pair of mirrors perpendicular to a drawing produces a circular repetition of the drawing. KaleidoMania!™ is an interactive software version of this idea.

In some amusement parks and science museums it is possible to find two large mirrors intersecting at a 90° angle (with the line of intersection parallel to the floor) so that when you look at yourself in the upper mirror you see your shoes, and in the lower mirror you see your head. It's an interesting experience for many people.

Three rectangular mirrors arranged to form the interior faces of a triangular prism make a kaleidoscope that can be placed vertically on a drawing or small object. When you look into the kaleidoscope from the open end you will see an "infinite triangular tiling" of the plane.

Using mirrored craft paper, it is easy to construct frustums of pyramids with interior mirrored faces. When the mirrored frustum is placed up side down on a table, we can see images of the regular polyhedra. In Figure II.20 we see the five frustums on the left, and a close-up of the one which produces the icosahedron on the right.

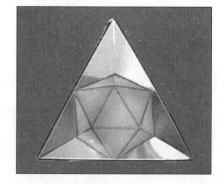

FIGURE II.20

In order to construct these special kaleidoscopes, the shapes of the quadrilateral sides (see Figure II.21) need to be determined. The quadrilaterals that make up the faces are isosceles trapezoids; hence it suffices to know the vertex angle of the original isosceles triangle in each case. This angle is the same as the angle at the center of the solid in question subtended by an edge of the solid.

The required vertex angle for each of the faces is: the tetrahedron, arccos({1/3} $\simeq$ 109°28′; the cube, arccos(1/3) $\simeq$ 70°32′; the octahedron, $\pi/2 = 90°$; the dodecahedron, arcsin(2/3) $\simeq$ 41°49′; and the icosahedron—arctan(2) $\simeq$ 63°26′.

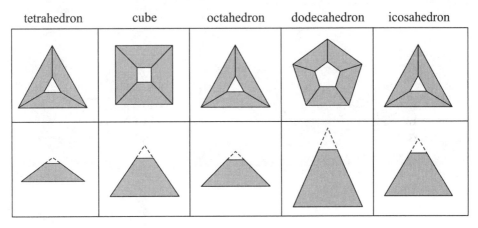

FIGURE **II.21**

Templates for constructing the sides of each frustum are presented in Figure II.22. In part (a) we have the template for the octahedron; and the rectangle in (b) whose sides are in the ratio of $\sqrt{2}$ to 1 will yield two pieces for the tetrahedron and two for the cube. Parts (c) and (d) yield pieces for the dodecahedron and the icosahedron, respectively.

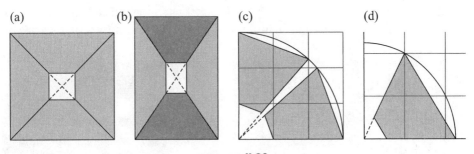

FIGURE **II.22**

Other effects are possible with triangular mirrors forming the interior faces of pyramids. Finally, if all six interior faces of a cube are mirrors, and a small pyramid is cut from one vertex to form a triangular window—what will be seen inside?

Concave or convex mirrors produce distorted images. For example, roll a rectangular sheet of silvered Mylar film (2 mils, i.e., 0.002 inch thick) into a cylinder, with the mirrored suface on the exterior. Images in a cylindrical mirror are called *anamorphic images*, or *anamorphs*. In Figure II.23 we see that an appropriately distorted planar image appears normal in the cylindrical mirror, and a template for distorting a rectangular image so that it will appear normal when the cylindrical mirror is place on the gray disk.

 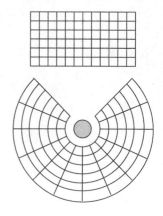

FIGURE **II.23**

Towards creativity

The Catalan architect Antoni Gaudí (1852–1926) was a very creative architect and investigator of new geometrical forms to be employed in architecture. He regularly used hands-on materials and produced scale models in expressing his remarkable three-dimensional creativity. In Figure II.24 we have photographs of his workshop in Barcelona.

His most important project was the Temple of the Holy Family (Sagrada Familia) which is still under construction in Barcelona. The two basic surfaces in this project are the hyperbolic paraboloid and the hyperboloid of one sheet. Since both surfaces are ruled surfaces, i.e., formed by straight lines, they are easy to construct, introducing an

FIGURE **II.24**

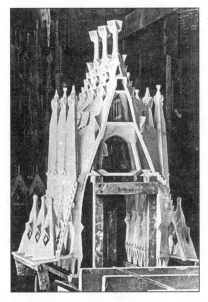 

FIGURE **II.25**

alternative to classical architectural forms. In Figure II.25 we see one of Gaudí's models for a portion of the Sagrada Familia, and a recent photograph of the actual building.

Our main reason for discussing Gaudí here is to recall some of his ideas on working visually and experimenting.

He developed a creative intuition that was ready for application...

> *"My structural and aesthetic ideas have a logic which cannot be denied. I have been thinking a lot why these ideas have not been applied before and this provokes my doubts. But since I am so convinced of their perfection, I have the obligation to apply them..."*

He rejected analytical approaches, and always worked directly from visualizations:

> *"Geometry simplifies constructions, algebraic expressions complicate them."*

Nature and real-world objects served as inspiration:

> *"This tree, near my workshop, is my master."*

After imagining a project, he paid great attention to all possible details:

> *"I carry out computations and experiments, I pay attention to all the details ...thus the logical shape is born from needs."*

In this way, he searched for equilibrium between rational and emotional characteristics:

> *"For a work to be beautiful, all its elements must be in the right location, must have the right dimension, the right shape, the right color ...to obtain harmony, you need contrast."*

On one occasion, Gaudí showed one of his visitors a range of geometrical models and, after describing their secrets and uses with emphatic use of his hands, he remarked, with excitement in his eyes:

"Wouldn't it be beautiful to learn geometry in this way?"

This challenge is as relevant today as it was in Gaudí's time.

Part III

Hints and Solutions to the Challenges

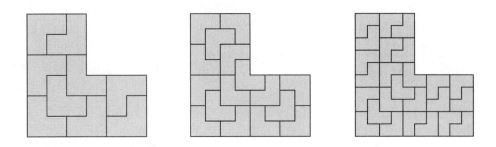

Many of the challenges in this book have multiple solutions. So here we give but one sample solution to each challenge, and encourage readers to search for others that may be simpler.

Chapter 1

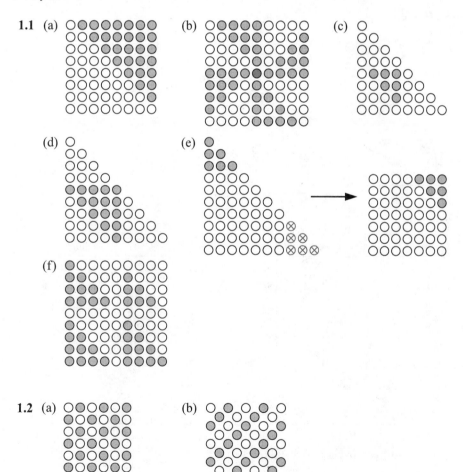

1.3

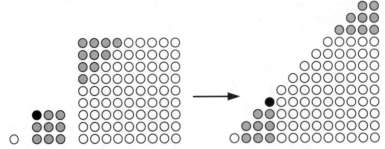

Chapter 2

2.1

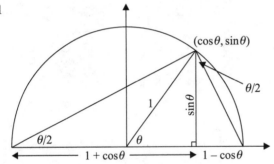

2.2

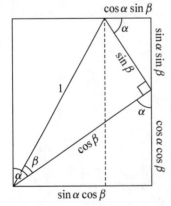

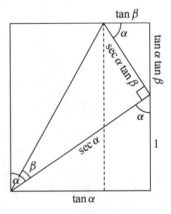

$$\sin(\alpha - \beta) = \sin\alpha \cos\beta - \cos\alpha \sin\beta$$
$$\cos(\alpha - \beta) = \cos\alpha \cos\beta + \sin\alpha \sin\beta$$

$$\tan(\alpha - \beta) = \frac{\tan\alpha - \tan\beta}{1 + \tan\alpha \tan\beta}$$

2.3 $\tan \gamma (\tan \alpha + \tan \beta) \ \tan \alpha \tan \beta$

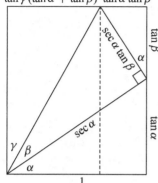

2.4 Using the diagram in the solution to Challenge 2.1, use the fact that the length of the arc subtended by the angle θ is at least as long as the hypotenuse of the triangle with legs $\sin \theta$ and $1 - \cos \theta$.

2.5 Sketch graphs of $y = (1 + x)^a$ and $y = 1 + ax$.

Chapter 3

3.1 $1 + 3 + 5 + \cdots + (2n - 1) = n^2$.

3.2

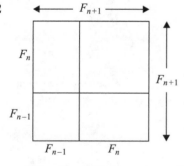

3.3

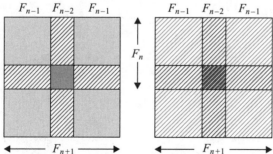

3.4 The resulting inequality is $1 \geq 4 \dfrac{ab}{(a + b)^2}$, which is equivalent to $\sqrt{ab} \geq \dfrac{2ab}{a + b}$.

3.5

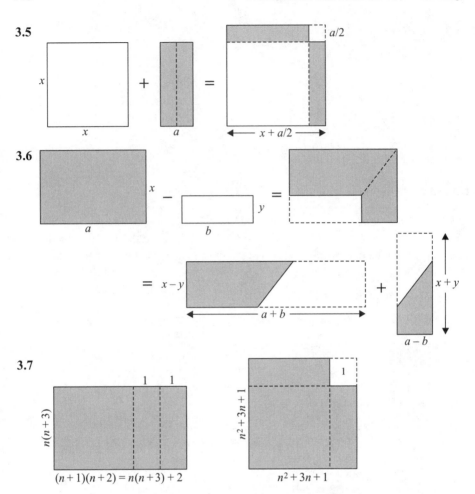

3.6

3.7

3.8 Note that the graph of $y = x^{p/q}$ partitions the unit square into two regions, and use integrals to evaluate the area of each part.

Chapter 4

4.1

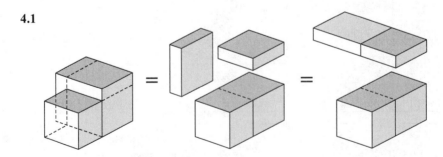

4.2
$$\sum_{i=1}^{m}\sum_{j=1}^{n}\left[a+(i-1)b+(j-1)c\right]=mna+n\frac{(m-1)m}{2}b+m\frac{(n-1)n}{2}c$$
$$=\frac{mn}{2}\left[2a+(m-1)b+(n-1)c\right]$$

4.3 Use the Fubini principle to count the cubes in two different ways.

Chapter 5

5.1

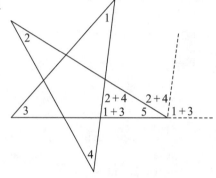

5.2 Let S denote the center of C_1, and draw the dashed lines as indicated at the right. Since $\overline{OP} = \overline{OQ} = \overline{OR}$, $\triangle OPQ$ and $\triangle OPR$ are isosceles. Let $\alpha = \angle PQO = \angle OPQ$ and $\beta = \angle PRO = \angle OPR$. Then $\angle PSO = 2\alpha$, and hence $\pi/2 = \angle SPR = (\pi/2 - 2\alpha) + \alpha + \beta$. Thus $\alpha = \beta$, $\triangle OPQ$ and $\triangle OPR$ are congruent, and $\overline{PQ} = \overline{PR}$.

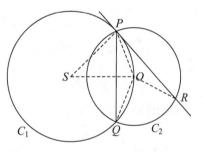

5.3

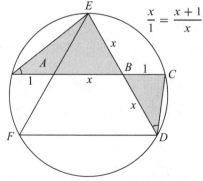

$$\frac{x}{1} = \frac{x+1}{x}$$

5.4 Yes.

Chapter 6

6.1 You need only translate the upper dark gray triangle and the left-hand side light gray triangle.

6.2 There are many proofs, this one is due to James Tanton [Tanton 2001b]:

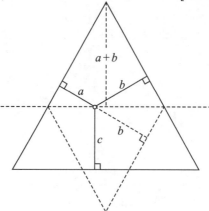

6.3 Inscribe the given octagon in a circle and divide it into triangles from the center. Since rotation around the center does not change areas, one can rearrange the triangles alternating the ones with base lengths 2 and 3. This octagon can be inscribed in a square of side $3 + 2\sqrt{2}$, and thus the area of the octagon is $(3 + 2\sqrt{2})^2 - 4(\sqrt{2} \cdot \sqrt{2})/2 = 13 + 2\sqrt{2}$.

6.4 In fact, more is true: each of the six angles at F (Figure 6.4) is 60°.

Chapter 7

7.1 In the figure below, use the fact that the shaded triangles are similar, and that $K = hc/2$.

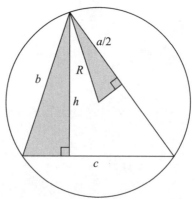

7.2 In Figure 7.3(c), bisect angle α in the gray triangle and bisect the 108° angle in the striped triangle.

7.3 Using the fact that $ab = ch$ (where c is the hypotenuse of the right triangle with sides a and b), show that a triangle with sides $1/a$, $1/b$, and $1/h$ is similar to the original triangle.

7.4 Connect the midpoints of the sides, and use the result in Section 7.4.

Chapter 8

8.1 Yes. Employ the procedure in the last paragraph of Section 8.2.

8.2 One answer is whenever ax and by have the same sign and $a/b = x/y$.

8.3 Yes. Placing a pencil mark on a cylinder of radius R and rolling the cylinder one revolution allows us to mark exactly a segment of length $2\pi R$. One could also use the graph of a cycloid.

8.4

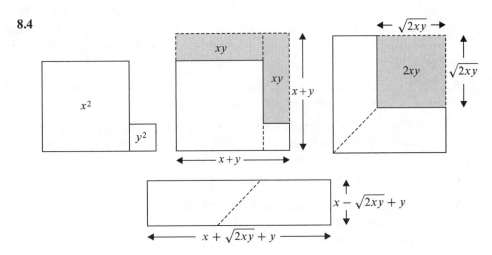

8.5 The areas of the parallelograms are $b \sin \alpha$ and $a \sin \beta$.

8.6 Show that $e + d + f = g + d + h$.

8.7

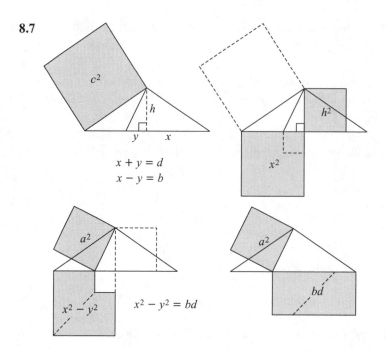

$$x + y = d$$
$$x - y = b$$

$$x^2 - y^2 = bd$$

Chapter 9

9.1 The problem has many solutions. For example, measure the perimeter p of the rectangular top, and starting from one of the corners mark points on the top corresponding to the five distances of length $p/5$. Mark the center of the top and cut the cake along the lines joining the center with the five division points on the perimeter of the top.

9.2 For a two-dimensional solution, partition a rectangle with dimensions $a + b$ and $(a + b)^2$; for a three-dimensional solution, partition a cube with edge length $a + b$.

9.3 Two distinct circles in the plane meet in (at most) two points. By rotating this figure in space about the line through the two centers one can visualize why the spheres intersect in a circle (the one described by the rotation of the two intersection points).

9.4 Follow the method presented in Section 9.3.

9.5 If you unfold the room as shown at the right, the distance from the spider to the fly is exactly 40 feet, and this is the shortest path.

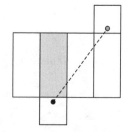

Chapter 10

10.1 The smaller square has $2/5$ ($= 4/10$) the area of the larger:

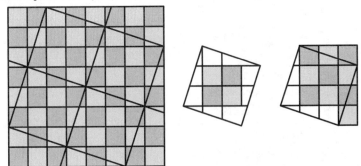

For the fraction k/n, the smaller square has $(n-k)^2/(k^2+n^2)$ the area of the larger.

10.2

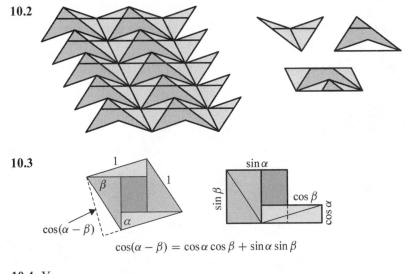

10.3

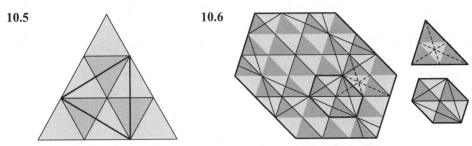

$$\cos(\alpha - \beta) = \cos\alpha\cos\beta + \sin\alpha\sin\beta$$

10.4 Yes.

10.5   **10.6**

10.7 If equilateral triangles are constructed outwardly on the sides of an arbitrary triangle, then their centers form the vertices of another equilateral triangle.

Chapter 11

11.1 Add three copies of the gray triangle to form a square as in the right-hand portion of Figure 6.1.

11.2

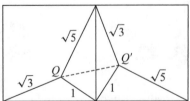

11.3 Arrange three copies (scaled to have height $b - a$) to form a cube of side length b with a smaller corner cube of side length a removed.

11.4 Re-label the lower right triangle so that α is the other (larger) acute angle, and use the identity $\cos(\alpha - \beta) = \sin\left[\pi/2 - (\alpha - \beta)\right]$.

Chapter 12

12.1 After the rotation, each gray triangle has the same base and height as the original triangle:

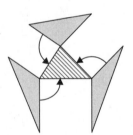
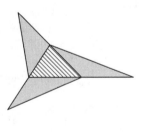

12.2 135°. The length of QQ' is $\sqrt{2}$, and the segments of lengths $\sqrt{2}$, $\sqrt{3}$ and $\sqrt{5}$ form a right triangle.

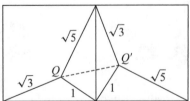

12.3 Consider several cases.

Chapter 13

13.1
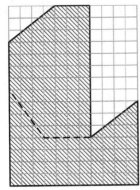
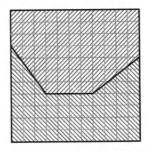

13.2 Assume the base and height of the concave pentagon are both 2, so that its area is 3. Then the dimensions of Loyd's "square" are 12/7 by 7/4, not $\sqrt{3}$ by $\sqrt{3}$.

13.3
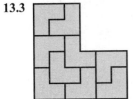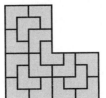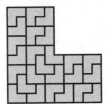

13.4
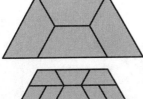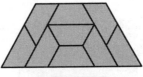
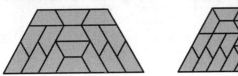

13.5 There are many solutions, here's but one:

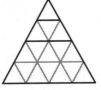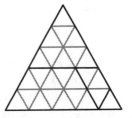

Chapter 14

14.1 a) Starting at $(a, f(a))$, move horizontally to $(f(a), f(a))$, vertically to $(f(a), f(f(a)))$, and so on.

b) Note that the sequence of points converges to $(2, 2)$.

14.2 For any $a \geq 0$, translate the graph of f by means of the vector $(a, f(a))$. If f is subadditive, this parallel curve will always lie above the graph of $y = f(x)$.

14.3 For continuous functions, convexity means that the region $\{(x, y) | y \geq f(x)\}$ is convex, i.e., any segment joining two points of the graph lies above the corresponding portion of the graph.

14.4 Move a copy of the axes, keeping the origin on the graph. The graph of $y = f(x)$ must always lie in the first and third quadrant for an increasing function, or in the second and fourth quadrant for a decreasing function.

Chapter 15

15.1 Many solutions are possible, here are two [Mabry, 2001]:

(a)   (b)

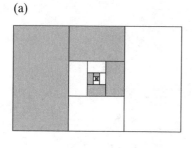 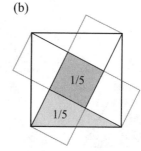 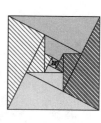

[For the left-hand part of (b), see the first Theorem in Section 10.2.]

15.2 $\frac{1}{8} + \frac{1}{16} + \frac{1}{32} + \cdots = \frac{1}{4}$ and $\frac{2}{9} + \frac{2}{81} + \frac{2}{729} + \cdots = \frac{1}{4}$.

15.3 Observe that $1 + 2(1 + 3 + 3^2 + \cdots + 3^n) = 3^{n+1}$.

15.4 In the coordinate plane, draw the rectangle $[0, \phi] \times [0, 1]$ and the line $y = x/\phi$. With compasses centered at $(\phi, 0)$ mark the point $(1 + \phi, 0)$ and recall that $1 + \phi = \phi^2$. Draw the vertical segment joining $(\phi^2, 0)$ to (ϕ^2, ϕ). Then with center $(\phi^2, 0)$ locate the point $(\phi^2 + \phi, 0)$ and recall that $\phi^2 + \phi = \phi(\phi + 1) = \phi^3$, and so on.

15.5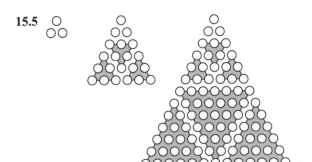

Chapter 16

16.1 a) Yes, since a 4×4 board can be covered as shown at the right:

b) No. The same argument given in Section 16.2 with the
standard coloring in Figure 16.2(c) will work.

16.2 a) There are 22 dark gray squares, 21 light gray and 21 medium gray squares in
each coloring, and since each straight tromino covers one square of each color, a
square which is dark gray in both colorings must be removed. Thus if the square
removed is not one of the four shown in (i) below, the board cannot be tiled. If the
square removed is one of the four shown, the board can be tiled, as shown in (ii)
below (many other tilings are possible).

(i)   (ii)

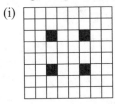

b) First consider the 4×4 and 5×5 boards. Then look for a relationship between
the $n \times n$ and $(n + 3) \times (n + 3)$ boards.

c) Tile three quadrants of the board as shown in (iii) below, and the other quadrant
with an appropriate reflection or rotation of one of the tilings in (iv) below.

(iii)   (iv)

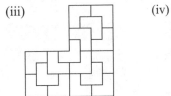   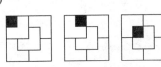

16.3 Proceed by induction on the number n of lines. Clearly two colors suffice when
$n = 1$. If a plane with n lines can be colored with two colors, so can one with an
additional line—simply reverse the colors on one side of the new line.

16.4 Two. Apply the same induction as in the preceding exercise.

16.5 Let $R(m, n)$ denote the $m \times n$ checkerboard. We will show that $R(m, n)$ may be covered with L-tetrominoes if and only if $m, n > 1$ and 8 divides mn. Assume that $R(m, n)$ can be tiled with L-tetrominoes, hence $m, n > 1$ and 4 divides mn, i.e., at least one of m and n must be even. Assume n is even and alternate black and white as colors for the n columns of $R(m, n)$. Let x denote the number of L-tetrominoes covering 3 black squares and 1 white square, and let y be the number covering 3 white and 1 black. Then $x + y = mn/4$ and $3x + y = x + 3y$, so that $x = y$ and $2x = mn/4$, hence 8 divides mn. The converse is immediate since it is easy to form 2×4 rectangles with two L-tetrominoes and 8×3 rectangles with six L-tetrominoes.

Chapter 17

17.1 Observe that the union of the two triangles contains a square of area 1.

17.2 Since the triangle OPA is included in the sector OPA we have the corresponding inequality of areas $(1 \cdot \sin x)/2 \le (\pi \cdot 1^2 \cdot x)/2\pi$, i.e., $\sin x \le x$. Moreover the inclusion of region PMA in sector PMB forces the arc PA to be shorter than the arc PB, so $x \le \pi \sin x/2$, i.e., $2x/\pi \le \sin x$.

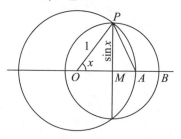

17.3 (a) Apply square roots to the equation in Figure 17.2.

17.4 In the first inequality, replace a and b by $|a|$ and $|b|$ and use the fact that $|a| + |b| \ge a + b$. The n-variables inequality follows similarly.

17.5

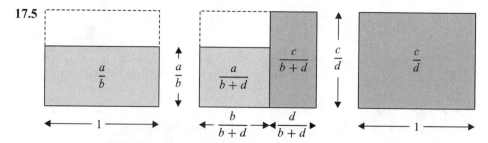

Chapter 18

18.1 Cutting the Möbius strips as suggested will produce more sophisticated strips (with more twists, with knots, or with a "chain" of loops).

18.2 *A* covers more than half the area of *B*—consider the area of the triangle whose vertices are the common corner of *A* and *B*, the right-most corner of *B*, and the left-most corner of *A*.

18.3 One volume is $\pi(a/2\pi)^2 b$ and the other is $\pi(b/2\pi)^2 a$, so the ratio of the volumes is the same as the ratio of side lengths.

18.4 Here's a solution due to
Ed Pegg, Jr. [Pegg, 2004]:

Chapter 19

19.1 There are 35 hexominoes. Eleven of them can be folded to form a cube.

19.2 Just two. Note that there will be one vertex of the die where faces 1, 2, and 3 meet, and these faces can surround the vertex clockwise or counterclockwise.

19.3 Consider pyramids, bipyramids, prisms, ... (you may wish to leave some faces blank).

19.4 At some vertices three faces meet, while at others four faces meet. We don't recommend it as a die!

19.5 Cube: triangles, squares, hexagon. Tetrahedron: triangles, square. Octahedron: squares, hexagon.

19.6 The solution to this problem consists of a rather large collection of curves. Perhaps the most surprising is the *lemniscate* (like the ∞ symbol). How do you cut the torus to produce this curve?

19.7 $\sqrt{3}/2$.

19.8 Since the solid is formed by joining two triangular prisms at their square bases, first show that space on one side of a plane can be filled by triangular prisms.

Chapter 20

20.1 Yes. For a vector proof, see [Kandall, 2002].

20.2 Yes. In fact, it is true if the vertices of the quadrilateral are any four non-collinear points in space.

20.3

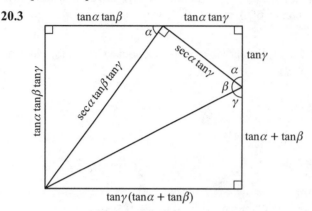

20.4 Draw the unit circle with center O, the line $ax+by = 0$, the point $P = (\cos t, \sin t)$ (so $\overline{OP} = 1$), and let Q be the orthogonal projection of P onto the given line. Since $\overline{PQ} \leq 1$, it follows that $|a\cos t + b\sin t|/\sqrt{a^2 + b^2} \leq 1$, from which we have $-\sqrt{a^2 + b^2} \leq |a\cos t + b\sin t| \leq \sqrt{a^2 + b^2}$ [Bayat et al. 2004].

20.5 Draw the triangle ABC with its three medians. With another copy of ABC, form a parallelogram with AB as one of the diagonals. Within this figure one can now locate a triangle whose sides are precisely two-thirds as long as the three medians. Reverse the steps to construct ABC given the medians.

20.6 On side $BC = a$ draw a semicircle of diameter a. Since the foot of the altitude h_b will lie on this semicircle (h_b is perpendicular to b), we can locate the foot of h_b and draw a line from C through the foot of h_b. Then A is the intersection of this line with a line parallel to BC at a distance of h_a.

20.7 In this case, the circular sector is at least a semicircle.

20.8 No:

(a)   (b)

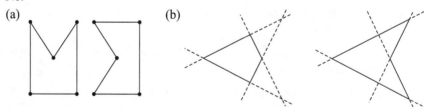

20.9 Good luck with this open problem!

References

S. A. Ajose, "Geometric series," *Mathematics Magazine*, vol. 67, no. 3 (June 1994), p. 230.

C. Alsina, "Neither a microscope nor a telescope, just a mathscope," in P. Galbraith et al. eds., *Mathematical Modelling: Teaching and Assessment in a Technology-Rich World*, Ellis Horwood, Chichester, 1998, pp. 3–10.

——, *Sorpresas Geométricas*, OMA, Buenos Aires, 2000a.

——, "The arithmetic mean-geometric mean inequality for three positive numbers," *Mathematics Magazine*, vol. 73, no. 2 (April 2000b), p. 97.

——, "Too much is not enough. Teaching maths through useful applications with local and global perspectives," *Educational Studies in Mathematics* 50, 2002, pp. 239–250.

——, "Cauchy-Schwarz inequality," *Mathematics Magazine*, vol. 77, no. 1 (February 2004), p. 30.

——, "Less chalk, less words, less symbols ... more objects, more context, more actions" in *ICMI Study 14: Applications and Modelling in Mathematics Education*. H.W. Henn and W. Blum eds., Springer, Berlin, 2006 (to appear).

——, "Mathematical proofs in the classroom: The role of images and hands-on materials," in *Mathematikunterricht im Spannungsfeld von Evolution und Evaluation— Festschrift für Werner Blum*, W. Henn and G. Kaiser, eds., Franzbecker, Hildesheim (2005) pp 129–138.

C. Alsina, C. Burgués, J.M. Fortuny, J. Giménez, J. and M. Torra, *Enseñar Matemáticas*, Editorial Graó, Barcelona 1996.

C. Alsina, and A. Monreal, "Proof without words: $(a + b)^3 = a^3 + 3a^2b + 3ab^2 + b^3$," *Teaching Mathematics and Computer Science* 1/1 (2003), p. 157.

——, "Proof without words: Beyond the parallelogram law," *Teaching Mathematics and Computer Science* 1/1 (2003) pp. 155–156.

Annairizi of Arabia, <http://tug.org/applications/PSTricks/Tilings>.

R. Arhneim, *Art and Visual Perception: A Psychology of the Creative Eye*, Faber and Faber, London, 1956.

——, *Visual Thinking*, Faber and Faber, London, 1970.

M. Bayat, M. Hassani, and H. Teimoori, "Proof without words: Extrema of the function $a \cos t + b \sin t$," *Mathematics Magazine*, vol. 77, no. 4 (October 2004), p. 259.

P. Beckmann, *A History of* π, St. Martin's Press, New York, 1971.

A. G. Bell, "The tetrahedral principle in kite structure," *National Geographic Magazine* 44 (1903) pp. 219–251.

M. Bicknell and V. E. Hoggatt Jr., eds., *A Primer for the Fibonacci Numbers*, The Fibonacci Association, San Jose, 1972.

M. Biermann, and W. Blum, "Realitäsbezogenes Beweisen-Der 'Schorle-Beweis' und andere Beispiele," *Mathematik Lehren* (110), 2002, pp. 19–22.

I. C. Bivens and B. G. Klein, "Geometric Series," *Mathematics Magazine*, vol. 61, no. 4 (October 1988), p. 219.

W. Blum and A. Kirsch, "Preformal proving: Examples and reflections," *Educational Studies in Mathematics*, 22 (2), 1991, pp. 183–203.

——, "Die beiden Hauptsätze der Differential- und Integralrechnung," *Mathematik Lehren* (78), 1996, pp. 60–65.

A. Bogomolny, Interactive Mathematics Miscellany and Puzzles, <http://www.cut-the-knot.org/content.shtml>, 1996.

B. Bolt, *Mathematics Meets Technology*, Cambridge, University Press, Cambridge, 1991.

M. Bosch, *La dimensión ostensiva en la actividad matemática. El caso de la proporcionalidad.* Tesis, Univ. Autonoma Barcelona, 1994.

C. B. Boyer, *A History of Mathematics*, John Wiley & Sons, New York, 1968.

A. Brousseau, "Sums of squares of Fibonacci numbers," in *A Primer for the Fibonacci Numbers*, M. Bicknell and V. E. Hoggatt Jr., eds. (1972), p. 147.

J. R. Brown, *Philosophy of Mathematics, An Introduction to the World of Proofs and Pictures*, Routledge, New York, 1999.

F. Burk, "The Pythagorean theorem," *College Mathematics Journal*, vol. 27, no. 5 (November 1996), p. 409.

B. Casselman, "Pictures and proofs," *Notices AMS*, November 2000, pp. 1257–1266.

J. H. Conway and R. Guy, *The Book of Numbers*, Copernicus, New York, 1996.

T. A. Cook, *The Curves of Life: Being an Account of Spiral Formations and their Applications to Growth in Nature, to Science, and to Art*, Dover Publications, New York, 1979.

H. S. M. Coxeter, *Introduction to Geometry*, Wiley, New York, 1969.

H. S. M. Coxeter and S.L. Greitzer, *Geometry Revisited*, MAA, Washington, 1967.

P. R. Cromwell, *Polyhedra*, Cambridge Univ. Press, Cambridge, 1999.

A. Cupillari, "Sums of cubes," *Mathematics Magazine*, vol. 62, no. 3 (October 1989), p. 259.

A. Cusmariu, "A proof of the arithmetic mean-geometric mean inequality," *The American Mathematical Monthly*, Vol. 88, No. 3. (March 1981), pp. 192–194.

P. J. Davis, "Visual theorems," *Educational Studies in Mathematics*, 24 (1993), pp. 333–344.

P. J. Davis and R. Hersh, *The Mathematical Experience*, Birkhäuser, Boston, 1981.

M. de Guzmán, *Para pensar major*. Ed. Labor, Barcelona, 1991.

——, *El rincón de la pizarra. Ensayos de visualización en Análisis Matemático*, Pirámide, Madrid, 1996.

M. De Villiers, "The role and function of proof in mathematics," *Pythagoras* 24, 1990, pp. 17–24.

——, *Rethinking Proof with Geometer's Sketchpad*. Key Curriculum Press, San Francisco, 1999.

——, "The value of experimentation in mathematics." *Proceedings 96th Nat. Cong. AMESA*, Cape Town, 2003, pp. 174–185.

——, "Developing understanding for different roles of proof in dynamic geometry," <http://mzone.mweb.co.za/residents/profmd/homepage. html>.

J. B. Dence and T. P. Dence, "A property of quadrilaterals," *College Mathematics Journal*, vol. 32, no. 4 (September 2001), pp. 291–294.

T. Dreyfus, "Imagery and reasoning in mathematics and mathematics education," *ICME-7* (1992) *Selected Lectures*, Les Presses Univ. Laval, Québec (1994), pp. 107–123.

R. H. Eddy, "A theorem about right triangles," *College Mathematics Journal*, vol. 22, no. 5 (November 1991), p. 420.

——, "Proof without words," *Mathematics Magazine*, vol. 65, no. 5 (December 1992), p. 356.

C. C. Edwards and P. S. Sonsgiry, "The distributive property of the triple scalar product," *Mathematics Magazine*, vol. 70, no. 2 (April 1997), p. 118.

J. Estalella, *Ciencia Recreativa*, Gustavo Gili, Barcelona, 1920.

H. Eves, *An Introduction to the History of Mathematics*, Holt, Rinehart, Winston, New York, 1976.

——, *Great Moments in Mathematics (before 1650)*, MAA, Washington, 1980.

S. J. Farlow, "Sums of integers," *College Mathematics Journal*, vol. 26, no. 3 (May 1995), p. 190.

J. Fauvel and J. van Maanen, eds., *History in Mathematics Education*. The ICMI Study, Kluwer Acad. Pub., Dordrecht, 2000.

A. Flores, "Tiling with squares and parallelograms," *College Mathematics Journal*, vol. 28, no. 3 (May 1997), p. 171.

G. N. Frederickson, *Dissections: Plane & Fancy*, Cambridge University Press, New York, 1997.

——, *Hinged Dissections: Swinging & Twisting*, Cambridge University Press, New York, 2002.

C. D. Gallant, "Proof without words: A truly geometric inequality," *Mathematics Magazine*, vol. 50, no. 2 (March 1977), p. 98.

——, "Proof without words: Comparing B^A and B^A for $0 < A < B$," *Mathematics Magazine*, vol. 64, no.1, (February 1991), p. 31.

G. Gamow, *One, Two, Three … Infinity*, Bantam Books, New York, 1961.

M. Gardner, *More Mathematical Puzzles and Diversions*, Penguin Books, Harmondsworth, U. K., 1961.

——, "Mathematical games," *Scientific American*, vol. 229, no. 4 (October 1973), p. 115.

Y. D. Gau, "Area of the parallelogram determined by vectors (a, b) and (c, d)," *Mathematics Magazine*, vol. 64, no. 5 (December 1991), p. 339.

G. Gheverghese Joseph, *La Cresta del Pavo Real. Las Matemáticas y sus raíces no europeas*. Pirámide, Madrid, 1996.

R. A. Gibbs, "The mediant property," *Mathematics Magazine*, vol. 63, no. 3 (June 1990), p. 172.

R. J. Gillings, *Mathematics in the Time of the Pharaohs*, The MIT Press, Cambridge, 1972.

J. Giménez, *100 imágenes ¡ 1000 problemas, pero ayudan a resolverlos*. Apuntes XXVI J.R.P.-OMA, Buenos Aires, 2001.

S. W. Golomb, "Tiling with trominoes," *American Mathematical Monthly*, vol. 81, no. 10 (December 1959), pp. 675–682.

——, "A geometric proof of a famous identity," *Mathematical Gazette*, vol. 49, no. 368 (May 1965), p. 199.

B. Grünbaum and G. C. Shephard, *Tilings and Patterns*, W. H. Freeman, San Francisco, 1986.

B. H. Gundlach, *Historical Topics for the Mathematics Classroom*, National Council of Teachers of Mathematics Inc., 1965.

J. Hadamard, *The Psychology of Invention in the Mathematical Field*, Dover, New York, 1954.

G. Hanna, "Some pedagogical aspects of proof," *Interchange* 21(1) (1990), pp. 6–13.

G. Hanna and H. N. Jahnke, "Proof and proving," in: *International Handbook of Mathematics Education*, A. J. Bishop et al. eds., Kluwer, Dordrecht, 1996, pp. 877–908.

——, "Arguments from physics in mathematical proofs: An educational perspective," *for the learning of mathematics*, 22(3) (2002), pp. 38–45.

——, "Proving and modelling," in *ICMI Study 14: Applications and Modelling in Mathematics Education*, H. W. Henn and W. Blum, eds. 2004, pp. 109–114.

D. W. Henderson, *Experiencing Geometry: On Plane and Sphere*, Prentice Hall, Upper Saddle River, NJ, 1996.

V. F. Hendricks, K. F. Jorgensen, P. Mancosu, and S. A. Pedersen, eds., *Visualization, Explanation and Reasoning Styles in Mathematics*, Kluwer, Dordrecht, 2003.

H. W. Henn and W. Blum, eds., *ICMI Study 14: Applications and Modelling in Mathematics Education*, Pre-conference Volume, Univ. Dortmund, 2004.

R. Hersh, "Proving is convincing and explaining," *Educational Studies in Mathematics*, 24(4) (1993), pp. 389–399.

R. Honsberger, *Mathematical Gems III*, MAA, Washington, 1985.

J. Horgan, "The death of proof," *Scientific American*, 269(4) (1993), pp. 92–103.

G. Howson, "Mathematics and common sense," *8th ICME—Selected Lectures*, Saem-Thales, Seville, 1998, pp. 257–269.

W. Johnston and J. Kennedy, "Heptasection of a triangle," *The Mathematics Teacher*, vol. 86, no. 3 (March 1993), p. 192.

D. Kalman, "Sums of squares," *College Mathematics Journal*, vol. 22, no. 2 (March 1991), p. 124.

G. A. Kandall, "Euler's theorem for generalized quadrilaterals," *College Mathematics Journal*, vol. 33, no. 5 (November 2002), pp. 403–404.

K. Kawasaki, "Proof without words: Viviani's theorem," *Mathematics Magazine*, vol. 78, no. 3 (June 2005), p. 213.

S. J. Kung, "The volume of a frustum of a square pyramid," *College Mathematics Journal*, vol. 27, no. 1 (January 1996), p. 32.

I. Lakatos, *Proofs and Refutations*, Cambridge Univ. Press, Cambridge, 1976.

J. E. Littlewood, *A Mathematician's Miscellany*, Methuen, London, 1953.

D. Logothetti, "Alternating sums of squares," *Mathematics Magazine*, vol. 60, no. 5 (December 1987), p. 291.

E. S. Loomis, *The Pythagorean Proposition*, National Council of Teachers of Mathematics, Inc., 1969.

W. Lushbaugh, "Sums of cubes," *Mathematical Gazette*, vol. 49, no. 368 (May, 1965), p. 200.

R. Mabry, "Proof without words," *Mathematics Magazine*, vol. 72, no. 1 (February 1999), p. 63.

——, "Mathematics without words," *College Mathematics Journal*, vol. 32, no. 1 (January 2001), p. 19.

J. Malkevitch, ed., *Geometry's Future*, COMAP, Lexington, 1991.

Yu. Manin, "Truth, rigour, and common sense" in *Truth in Mathematics*, H. G. Dales & G. Oliveri (eds.), Oxford Univ. Press, Oxford, 1998, pp. 147–159.

G. E. Martin, *Polyominoes. A Guide to Puzzles and Problems in Tiling*, MAA, Washington, 1991.

J. H. Mason, *Mathematics Teaching Practice. A Guide for University and College Lectures*, Horwood Pub. Chichester, 2004.

K. Menger, *Calculus: A Modern Approach*, Ginn and Company, 1952.

F. Nakhli, "The vertex angles of a star sum to 180°," *College Mathematics Journal*, vol. 17, no. 4 (September 1986) p. 338.

NCTM, *Principles and Standards for School Mathematics*, National Council of Teachers of Mathematics Inc., 2000.

R. B. Nelsen, "The harmonic mean-geometric mean-arithmetic mean-root mean square inequality," *Mathematics Magazine*, vol. 60, no. 3 (June 1987) p. 158.

——, "Sums of cubes," *Mathematics Magazine*, vol. 63, no. 3 (June 1990) p. 178.

——, *Proofs without Words: Exercises in Visual Thinking*, MAA, Washington, 1993.(In Spanish: Proyecto Sur, Granada, 2001).

——, "The sum of a positive number and its reciprocal is at least two (four proofs)," *Mathematics Magazine*, vol. 67, no. 5 (December 1994), p. 374.

——, "Volume of a frustum of a square pyramid," *Mathematics Magazine*, vol. 68, no. 2 (April 1995), p. 109.

——, "The sum of the squares of consecutive triangular numbers is triangular," *Mathematics Magazine*, vol. 70, no. 2 (April 1997), p. 130.

——, "One figure, six identities," *College Mathematics Journal*, vol. 30, no. 5 (Nov. 1999), p. 433; vol. 31, no. 2 (March 2000a), pp. 145–146.

——, *Proofs without Words II: More Exercises in Visual Thinking*, MAA, Washington, 2000b.

——, "Heron's formula via proofs without words," *College Mathematics Journal*, vol. 34, no. 4 (September 2001), pp. 290–292.

——, "Paintings, plane tilings, & proofs," *Math Horizons*, MAA (November 2003), pp. 5–8.

——, "Proof without words: Four squares with constant area," *Mathematics Magazine*, vol. 77, n. 2 (April 2004), p. 135.

W. Page, "Geometric sums," *Mathematics Magazine*, vol. 54, no. 4 (September. 1981), p. 201.

E. Pegg Jr., 2004, <www.mathpuzzle.com>.

G. Pólya, *Mathematics and Plausible Reasoning: Induction and Analogy in Mathematics.* Vol. I, Princeton University Press, Princeton, 1954.

——, *Mathematical Discovery: On Understanding, Learning and Teaching Problem Solving* (2 vols., combined ed.) John Wiley & Sons, New York, 1981.

V. V. Prasolov, *Essays on Numbers and Figures*, American Mathematical Society, 2000.

V. Priebe and E. A. Ramos, "Proof without words: The sine of a sum," *Mathematics Magazine*, vol. 73, no. 5 (December 2000), p. 392.

J. Rabinow, *Inventing for Fun and Profit*, San Francisco Press, 1990.

Y. Rav, "Why do we prove theorems?" *Philosophia Mathematica*, 7(3) (1999), pp. 5–41.

J. W. S. Rayleigh, *Scientific Papers*, Dover Publications, New York, 1964.

P. R. Richard, *Raisonnennent et stratégies de preuve dans l'enseignement des mathématiques*, Peter Long, Berne, 2004.

I. Richards, "Sums of integers," *Mathematics Magazine*, vol. 57, no.2 (March 1984), p. 104.

W. Romaine, "Proof without words: $(\tan \theta + 1)^2 + (\cot \theta + 1)^2 = (\sec \theta + \csc \theta)^2$," *Mathematics Magazine*, vol. 61, no. 2 (April 1988), p. 113.

J. Rotman, *Journey into Mathematics: An Introduction to Proofs.* Prentice Hall, New York, 1998.

T. A. Romberg and J. de Lange, eds., *Mathematics in Context*, EBEC, Chicago, 1997.

G. C. Rota, "The phenomenology of mathematical proof," *Synthese*, 3(2) (1997), pp. 183–197.

N. Sanford, "Dividing a frosted cake," *Mathematics Magazine*, vol. 75, no. 4 (October 2002), p. 283.

L. Santaló, *La geometría en la formación de profesores*, OMA, Buenos Aires, 1993.

D. Schattschneider, "Proof without words: The arithmetic mean-geometric mean inequality," *Mathematics Magazine*, vol. 59, no. 1 (February 1986), p. 11.

G. Schrage, "Sums of integers and sums of cubes," *Mathematics Magazine*, vol. 65, no. 3 (June 1992), p. 185.

B. Schweizer, A. Sklar, K. Sigmund, P. Gruber, E. Hlawka, L. Reich, and L. Schmetterer, eds., *Karl Menger Selecta Mathematica* Volume 2, Springer, Vienna, 2003.

M. Senechal, "Visualization and visual thinking," in: *Geometry's Future*, J. Malkevitch, ed., (1991), pp 15–22.

M. Senechal and G. Fleck, *Shaping Space: A Polyheral Approach*, Design Science Collection, Birkhäuser, Boston, 1988.

D. B. Sher, "Sums of powers of three," *Mathematics and Computer Education*, vol. 31, no. 2 (Spring 1997), p. 190.

S. J. Shin, *The Logical Status of Diagrams*, Cambridge, UP, Cambridge, 1994.

A. Sierpinska, *Understanding in Mathematics*. Falmer, London, 1994.

M. K. Siu, " Sums of squares," *Mathematics Magazine*, vol. 57, no. 2 (March 1984), p. 92.

L. A. Steen, *For All Practical Purposes*, COMAP, Lexington. W. H. Freeman Co. New York, 1994.

S. K. Stein, "Existence out of chaos," in R. Honsberger, *Mathematical Plums*, 62–93, MAA, Washington, 1979.

H. Steinhaus, *Mathematical Snapshots*, G. E. Steichert & Co. New York, 1938.

P. D. Straffin, "Liu Hui and the first golden age of Chinese mathematics," *Mathematics Magazine*, vol. 71, no. 3 (June 1998) p. 170.

J. Tanton, *Solve This: Math Activities for Students and Clubs*, MAA, Washington, 2001a.

——, "Equilateral triangle," *Mathematics Magazine*, vol. 74, no. 4, (October 2001b), p. 313.

The Mathematics Initiative (EDC), "Cramer's rule," *College Mathematics Journal*, vol. 28, no. 2 (March 1997), p. 118.

J. van de Craats, "A golden section problem from the *Monthly*," *American Mathematical Monthly*, vol. 93, no. 7 (August-September 1986), p. 572.

E. Veloso, *Geometría*, APM, Lisboa, 2000.

J. H. Webb, "Geometric series," *Mathematics Magazine*, vol. 60, no. 3 (June 1987), p. 177.

E. Wittmann, "Operative proofs in primary mathematics," in: *Proofs and Proving: Why, When and How?* Proceedings of TG 8 at ICME-8 Sevilla, M. de Villiers, ed. Centrahil AMESA, 1996, pp. 15–22.

S. Wolf, "Viviani's theorem," *Mathematics Magazine*, vol. 62, no. 3 (June 1989), p. 190.

K. Y. Wong, *Using Multi-Modal Think-Board to Teach Mathematics*, TSG14/ ICME10, Copenhagen, 2004.

F. Yuefeng, "Jordan's inequality," *Mathematics Magazine*, vol. 69, no. 2 (April 1996), p. 126.

ZDM, *Analyses: Visualization in Mathematics and Didactics of Mathematics*, ZDM, Karlsruhe, 1994.

W. Zimmermann and S. Cunningham eds., *Visualization in Teaching and Learning Mathematics*, Notes 19, MAA, 1991.

Index

About the Authors

Claudi Alsina was born on 30 January 1952 in Barcelona, Spain. He received his BA and PhD in mathematics from the University of Barcelona. His post-doctoral studies were at the University of Massachusetts, Amherst. Claudi, Professor of mathematics at the Technical University of Catalonia, has developed a wide range of international activities, research papers, and publications in mathematics and mathematics education.

Roger B. Nelsen was born on 20 December 1942 in Chicago, Illinois. He received his BA in mathematics from DePauw University in 1964 and his PhD in mathematics from Duke University in 1969. Roger was elected to Phi Beta Kappa and Sigma Xi. His previous books include *Proofs Without Words: Exercises in Visual Thinking*, MAA, 1993; *An Introduction to Copulas*, Springer, 1999 (2nd ed. 2006); *Proofs Without Words II: More Exercises in Visual Thinking*, MAA, 2000.